RESEARCH REPORTS
OF THE LINK ENERGY FELLOWS

VOLUME 8

RESEARCH REPORTS OF THE LINK ENERGY FELLOWS

VOLUME 8

BRIAN J. THOMPSON, EDITOR

Published by
The University of Rochester Press
in association with
The Link Foundation

First published in 1993

University of Rochester Press
200 Administration Building, University of Rochester
Rochester, New York 14627, USA
and at PO Box 9, Woodbridge, Suffolk IP12 3DF, UK

ISBN 1 878822 20 9

Library of Congress Cataloging-in-Publication Data applied for

British Library Cataloguing-in-Publication Data
A catalogue record for this book is available from the British Library

This publication is printed on acid-free paper

Printed in The United States of America

TABLE OF CONTENTS

THE LINK FOUNDATION
ENERGY FELLOWSHIP PROGRAM

The Link Foundation was established in 1953 by Mr and Mrs Edwin A. Link. The purpose of the Link Foundation is to promote the general welfare through the advancement of scientific, technological and general educational projects. The University of Rochester administers the Link Foundation Energy Fellowships. Objectives of this program are to foster energy research and to disseminate results of that research through lectures, seminars and publication. Since 1984 grants have been made annually to universities or other non-profit organizations which select Doctoral student fellows and supervising faculty or research directors to pursue an energy-related project. Ideas that can be implemented in the relatively near future are given priority by an independent selection committee.

The volumes in the series of Link Energy Fellows Reports and of The Link Energy Conferences are the result of research papers prepared by the fellows following their year of research under the auspices of The Link Foundation. The Fellows are all students accepted in a Ph.D. program at their respective institutions and were chosen in a national competition. The papers represent interesting, if diverse, pieces of research that we expect will make stimulating reading as well as being a formal record of the work carried out under the program.

The Link Energy Fellowship Program is funded by the Link Foundation to both honor and memorialize Edwin A. Link. The Link Foundation was established in 1953 by Mr. and Mrs. Edwin A. Link. It is the policy of the Foundation to make grants to qualified non-profit organizations interested in the mastery of the air and sea, and the development of energy resources and their conservation. Each year these fellowships have been awarded in these categories as a means of implementing this policy.

The Link Energy Fellows were selected by an independent review committee from a large number of applications, and hence they represent some of the best young scholars in the energy and energy-related fields.

SELECTION COMMITTEE FOR 1992 FELLOWS

Provost Brian J. Thompson (Chair)
University of Rochester
Rochester, New York 14627

Dean Bruce W. Arden
College of Engineering and Applied Science
University of Rochester
Rochester, New York 14627

Professor William N. Gill
Chair, Department of Chemical Engineering
Rensselaer Polytechnic Institute
Troy, New York 12180-3590

Provost Angel G. Jordan
Carnegie Mellon University
5000 Forbes Avenue
Pittsburgh, Pennsylvania 15213

Professor Lee R. Lynd
Biotechnology and Biochemical Engineering Program
Thayer School of Engineering
Dartmouth College
Hanover, New Hampshire 03755

Sun-pumped Lasers:
Revisiting an Old Problem with Nonimaging Optics

DAVE COOKE

The University of Chicago
Department of Physics and The Enrico Fermi Institute
5640 S. Ellis Avenue
Chicago, IL 60637 U.S.A.

ABSTRACT

The techniques of nonimaging optics have allowed for the production of a world-record intensity of sunlight: 72 watts per square millimeter (W/mm^2), using a sapphire concentrator. Such an intensity exceeds the intensity of light at the surface of the sun itself (63 W/mm^2) by 15 percent and may have useful applications in pumping lasers, which require high intensities of light to function. This paper describes the production of high-intensity sunlight and reports its application in generating over three watts of laser power from a 72.5-centimeter-diameter telescope mirror at an efficiency exceeding that typically attained in approaches not involving nonimaging optics.

BACKGROUND

Shortly after the invention of the laser, investigators began dreaming of directly converting sunlight, an incoherent, broadband source, into laser radiation, a coherent, monochromatic source. The dreams were quickly realized [1, 2], but the efficiencies of sun-pumped lasers typically have been restricted to less than a percent. The chief reason for the low performance is that lasers have high thresholds: the amplifying medium must contain a high power per unit volume before any lasing can take place.

Nonimaging optics, a relatively new subdiscipline of optics, provides a means of concentrating light to intensities approaching the theoretical limit. Specifically, phase space conservation [3] and thermodynamic arguments [4] place a limit on the concentration of sunlight any optical device can attain, namely

$$C_{max} = n^2/\sin^2\theta, \tag{1}$$

where n is the index of refraction at the target surface and θ is the semiangle subtended by the sun. At the surface of the earth $\theta = 0.27^\circ$, and so with an ordinary refractive material (n $\sim$ 1.5) one obtains an upper limit on concentration of about 100,000.

The concentrations attained with conventional imaging devices fall far short of this limit, because of abberations [5]. A parabolic mirror, for example, produces a perfect image on axis, but blurs and broadens the image off axis. To attain high concentrations the requirement of forming images is relaxed. A laser crystal or a solar furnace does not care about receiving a picture-perfect image of the sun; all either cares about is receiving the maximum power per area. By dispensing with image-forming requirements in applications where no image is required, concentrations that approach the theoretical limit can be attained.

Nonimaging optics provides a framework for designing such powerful concentrators. A nonimaging concentrator is essentially a "funnel" for light. The chief approach to designing such a concentrator is called the "edge ray" method [5], in which all light rays entering the device at the maximum desired collection angle are directed after one reflection at most to the rim of the exit aperture (Figure 1). In this way all other rays inside the maximum acceptance angle are reflected within the aperture itself.

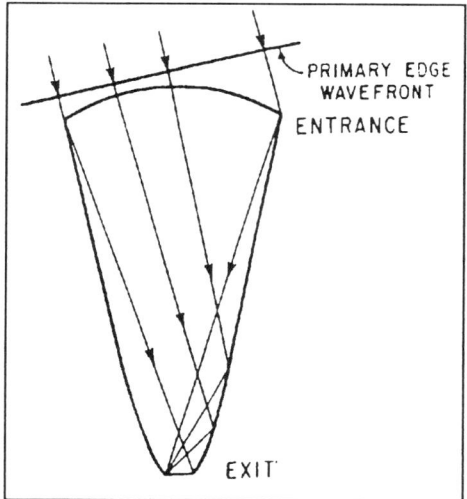

Figure 1. In a nonimaging concentrator designed by the edge ray method all light rays entering the device at the maximum collection angle are directed after one reflection at most to the rim of the exit aperture.

PRODUCING HIGH-INTENSITY SUNLIGHT

In principle high concentrations of sunlight could be attained with a nonimaging concentrator alone, but such a device would be quite large and unwieldy. A more practical approach for concentrating sunlight utilizes a two-stage system incorporating a focusing first stage primary mirror and a nonimaging second stage concentrator (Figure 2). The focusing first stage can be a parabolic or a spherical mirror. The overall concentration for a two-stage system is simply the product of the concentration of the primary mirror with that of the nonimaging concentrator:

$$C = n^2\cos^2\phi/\sin^2\theta, \tag{2}$$

which falls short of the theoretical limit (equation (1)) only by the factor of $\cos^2\phi$, where ϕ is the rim angle of the mirror (given approximately by the inverse tangent of the radius of the mirror, $D/2$, divided by its focal length, F). The effects can be minimized by making the rim angle small; i.e., choosing a mirror with a relatively large F/D ratio, or f number.

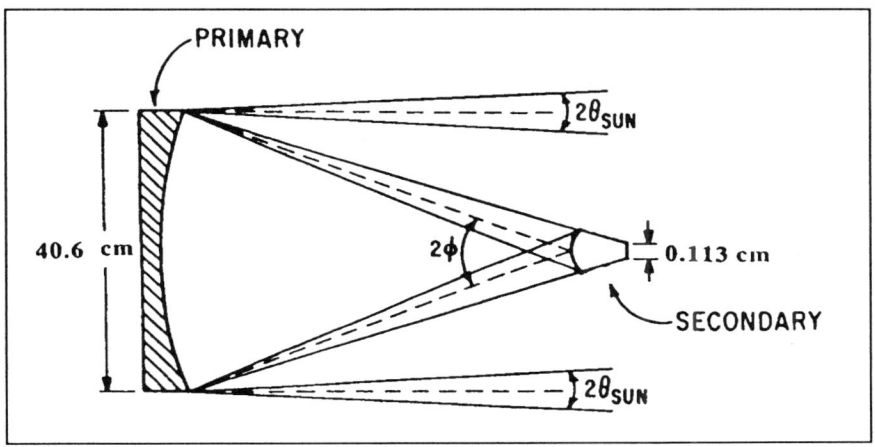

Figure 2. Two-stage approach to concentrating sunlight is illustrated.

Employing a technique based on this concept, in 1988 a vessel filled with an oil having an index of refraction of 1.53 was used to achieve a record intensity of sunlight of 44 W/mm^2 [6]. In 1990 a sapphire concentrator (n = 1.76) was employed to boost the intensity to 72 W/mm^2, which is 15 percent greater than the intensity of light at the surface of the sun itself. [7] (Sapphire was chosen for its high index of refraction and low absorption.) Besides increasing the index of refraction, the design of the concentrator was modified so that it worked by total internal reflection [8]. (In the oil-filled design the walls of the concentrator were silvered, and as such there were losses due to absorption. Total internal reflection, on the other hand, is nearly loss-free, and thereby increases the efficiency of the concentrator, allowing a greater fraction of the theoretical limit to be attained.)

To attain the world-record intensity of sunlight, a 40.6-cm-diameter silver coated telescope mirror with rim angle ϕ = 11.5° (f = 2.5) was used (Table I). Sunlight was concentrated to a one centimeter spot one meter away from the mirror. The light was further compressed to a one millimeter spot by the nonimaging sapphire concentrator. Substituting the relevant values into equation (2) gives the theoretical limit to the concentration the two-stage system:

$$C = [1.76\cos(11.5°)/\sin(0.27°)]^2 = 137{,}000. \qquad (3)$$

Table I

Two-Stage Concentrator Design Specifications

Primary

Diameter	40.6 cm
Focal length	101.6 cm
Rim angle	11.5°
Concentration ratio	1,730

Secondary

Entrance diameter	0.977 cm
Exit diameter	0.113 cm
Acceptance angle	11.5°
Refractive index	1.76
Concentration ratio	74.8

TOTAL CONCENTRATION = 129,000
(5.8% less than thermodynamic limit)

Measuring the actual concentration required some ingenuity. The intensity of the light exiting the tip of the sapphire concentrator would damage conventional power meters. Moreover, the light sprays out in a wide cone (filling nearly 2π steradian hemisphere), and it is contained in a dielectric. To surmount these difficulties the 19th century technique of calorimetry was relied upon (Figure 3). Light exiting the tip of the concentrator passed through a sapphire window and into a thermos bottle filled with a Cargille index matching fluid ($n = 1.76$). The tip of the concentrator was optically coupled to the sapphire window with a drop of a second, different matching fluid, tin chloride in glycerol. Sunlight entered the calorimeter for a 30-second interval. Three thermocouples inside the calorimeter permitted the recording of the equilibrium temperature of the liquid before and after each 30-second interval, from which the change in temperature ΔT_S could be determined (Figure 4). Then a heater inside the calorimeter of known power P_H was turned on for a 30-second interval, and the change in temperature ΔT_H was measured (Figure 5). The power delivered by the sun through the tip of the concentrator during a 30-second interval is therefore given by

$$P_S = (\Delta T_S / \Delta T_H) P_H \qquad (4)$$

A separate experiment was done to verify that all the sunlight entering the calorimeter does indeed pass through the tip. It was found that 3.8 percent of the light does not come from the tip. This leakage probably comes from the small amount of tin chloride solution spilling over at the tip.

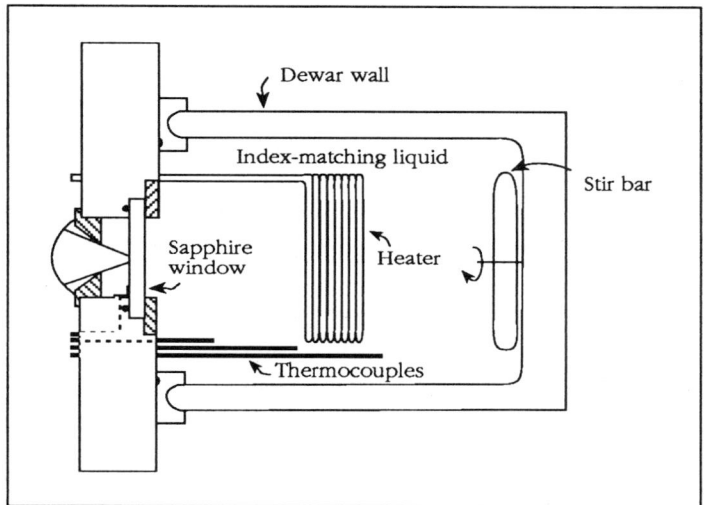

Figure 3.   Calorimeter, or thermos bottle, traps sunlight passing through the tip of the concentrator. The temperature rise of the liquid inside is calibrated with the electric heater.

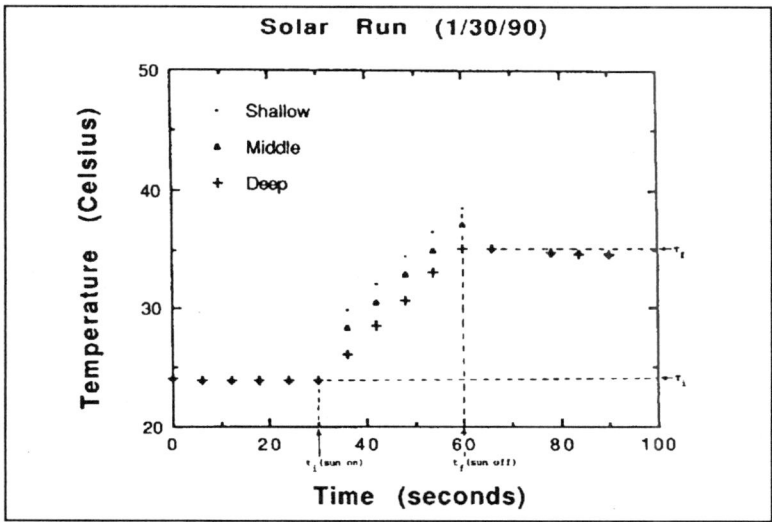

Figure 4.   Plot of temperature versus time during solar heating.

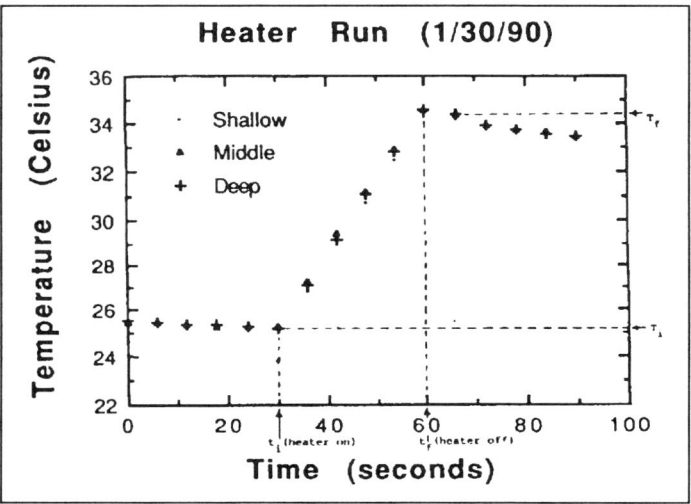

Figure 5.   Plot of temperature versus time during electrical heating.

In a series of trials over several cold, crisp days in January and February, 1990, an average power of 72 watts of sunlight was measured passing into the calorimeter, taking into account the 3.8 percent leakage. Dividing by the area of the tip, 1.0 mm^2, gives an intensity of 72 W/mm^2, which is 84,000 times greater than the direct incident intensity, 0.86 mW/mm^2. (The measured value falls short of the theoretical limit primarily because of losses at the reflecting surface of the primary mirror and blockage from the calorimeter.) (Table II) The light passing through the 1 millimeter tip of the concentrator is the highest intensity of sunlight anywhere in the universe, including at the surface of the sun itself (Table III).

Table II

Irradiance Loss Analysis

95,000 suns expected
(73.3% of geometrical limit)

Primary Losses

Reflection	5%
Detector shading	10%

Net Primary Efficiency	85.5%

Secondary

Lens reflection	8%
Skew rays	3%
Absorption	1%
Leakage	3%

Net Secondary Efficiency	85.7%

OVERALL EFFICIENCY = 73.3%

Table III

High Flux Data Summary

Date	NIP (W/m^2)	ΔT_s (°C)	ΔT_H (°C)	P_H (W)	P_S (W)	P/A (kW/cm^2)	Conc.
1/28/90	860	12.8	9.1	53.6	75.4	7.5	87,000
	870	12.5	9.2	52.8	71.7	7.2	82,000
	860	13.4	8.9	53.4	80.4	8.0	93,000
1/30/90	730	10.3	9.3	52.8	58.5	5.8	80,000
	710	11.0	9.0	53.1	64.9	6.5	91,000
2/20/90	840	12.3	8.7	53.5	75.6	7.5	90,000
	870	11.9	8.7	53.5	73.2	7.3	84,000

AVERAGE CONCENTRATION = 87,000 ± 3,000

CORRECTING FOR LEAKAGE THROUGH THE SIDES OF THE CONCENTRATOR YIELDS:

84,000 ± 3,000

APPLICATION OF NONIMAGING OPTICS TO LASER PUMPING

Having attained the highest concentrations of sunlight ever made, it seemed interesting to revisit the sun-pumped laser and see if the techniques of nonimaging optics would enable a greater efficiency of operation. The design of the laser system (Figure 6) follows the basic two-stage approach outlined above. The solar laboratory at the University of Chicago is situated on the roof of the high-energy physics building on campus. A one-meter-square heliostat redirects solar radiation to a small primary reflector. The reflective surface of the heliostat mirror consists of an enhanced aluminum front-surface coating called BV2, prepared by the Optical Coating Laboratories Incorporated (OCLI); using a normal-incidence pyroheliometer, the reflectivity of the mirror was measured to be 95 percent. The primary reflector is a front-surface silvered 72.5-centimeter-diameter telescope mirror having a large enough focal ratio (f/2.0) so that it could be made spherical instead of parabolic, with little loss of concentration--and at considerable less expense.

With an incident intensity of 900 W/m^2, the combined system of the heliostat and the primary mirror delivers about 200 watts of power--in the form of a spot approximately 1.5 centimeters in diameter--to the aperture of a nonimaging concentrator. (The power delivered to the concentrator was measured with a power meter.) The light from the primary is then further concentrated by the nonimaging concentrator.

The concentrator is made of fused silica, which has an index of refraction of about 1.5, and it "compresses" the 1.5-centimeter spot from the mirror to a 2.5-millimeter-diameter spot. One way of thinking about the laser system, then, is that the light reflected from the 72.5-centimeter-diameter mirror is "funnelled" into a 2.5-millimeter-diameter spot. The intensity of light emerging from the tip of the concentrator is about 40 W/mm^2 (=200 W/ $\pi(1.25mm)^2$). Such an intensity can also be calculated by multiplying the geometric ratio of the area of the mirror to the area of the concentrator tip, 84,000, by the efficiency of the heliostat mirror, 95 percent, and the combined efficiency of the primary and secondary, 60 percent. (The combined efficiency was determined in earlier work; it includes the effects of circumsolar radiation.) Multiplying the resulting concentration ratio of 45,000 by an incident flux of 900 W/m^2 yields the 40 W/mm^2 calculated above.

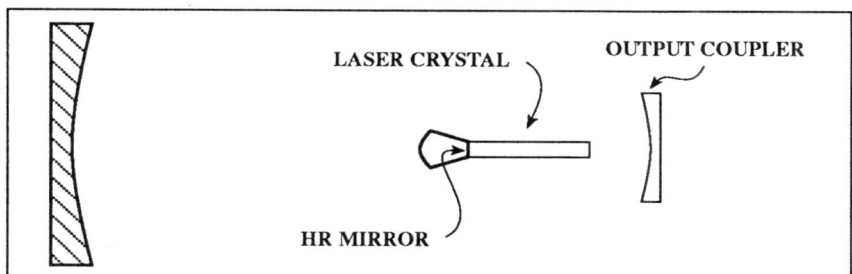

Figure 6.   Solar laser system.

The light exiting the tip of the concentrator is then directed into one end of a laser crystal; such an approach is called end pumping. (When the pumping light enters the side of the crystal, the approach is called side pumping.) The tip of the concentrator is optically coupled to the crystal with a small dab of Dow Corning optical couplant. To keep the couplant from overheating and burning, a spinning wheel with a hole in it is placed in front of the concentrator to chop the sunlight. The diameter of the hole is about 1/20th that of the wheel. A filter having a cut-off at 495 nanometers is positioned between the spinning wheel and the concentrator to shield the laser crystal from the ultraviolet portion of the solar spectrum.

Measurements were made for two types of crystals, Nd:YAG (neodymium-doped yttrium aluminum garnet) and Nd:Cr:GSGG (neodymium- and chromium-doped gadolinium scandium gallium garnet), both of which were supplied by Litton Airtron. The active lasing element in both crystals is neodymium (1.6 atomic percent in Nd:YAG and 2.4 atomic percent in Nd:-Cr:GSGG), which has an output at 1.06 microns. Chromium (1.2 atomic percent) is employed in the GSGG host matrix as a "cascade" to absorb sunlight at frequencies not absorbed by the neodymium and transfer that energy to it. There were four crystals in all--two lengths of each type, 10 millimeters and 20 millimeters. The barrel diameters of all four crystals were 2.5 millimeters. The coupled end of the laser rod was coated with a high-reflection (HR) film at 1.06 microns, and so the HR coated end served as the end mirror of the laser cavity. The other end of the crystal was anti-reflection (AR) coated at 1.06 microns. An output coupler, or external partially transmitting mirror, completed the laser cavity.

OUTPUT OF SUN-PUMPED LASERS

Both types of crystals were successfully lased. The chopping wheel allowed for the production of laser pulses having a duration of approximately 3 milliseconds. Three 2-meter-radius-of-curvature output couplers, having reflectivities of 99, 95 and 90 percent, were used. The 95 percent reflective mirror yielded the best results. The laser output was measured as a function of the input power (Figure 7), which was varied by masking the primary mirror with pie-shaped pieces of non-reflecting metal. Here, the input power is taken to mean the power delivered to the nonimaging concentrator by the heliostat and primary mirror. (The input power is approximately 55 percent of the product of the area of the primary mirror with the incident solar flux.) Up to 45 percent of the total area of the mirror could be masked, at intervals of 5 percent. The detector employed to measure the laser output was calibrated with a laboratory continuous-wave (CW) Nd:YAG laser, having an output power that could be varied from about 1 to 18 watts.

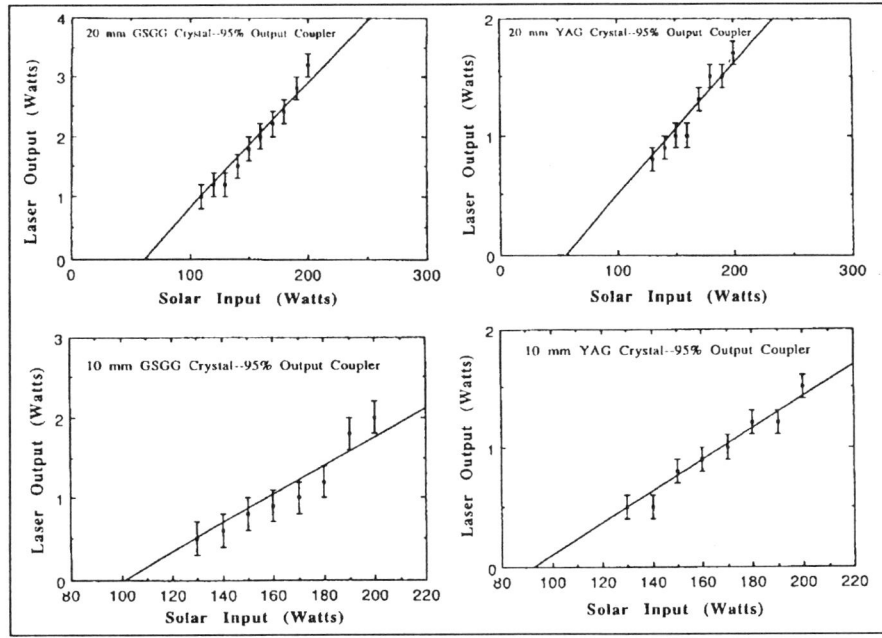

Figure 7. Plots of laser output versus solar input

The best results were obtained with the 20-millimeter-long Nd:Cr:GSGG crystal, which yielded 3.2 watts of power, for an overall efficiency of more than 1.5 percent (3.2 watts/200 watts) and a slope-efficiency (the efficiency above threshold) of greater than 2 percent. Threshold was determined to be approximately 60 watts of incident power, which corresponds to an intensity of about 10 W/mm^2 incident on the end of the laser crystal. The 20-millimeter-long Nd:YAG crystal had an overall efficiency of slightly less than 1 percent (1.7 watts/200 watts) and a slope-efficiency somewhat greater than 1 percent.

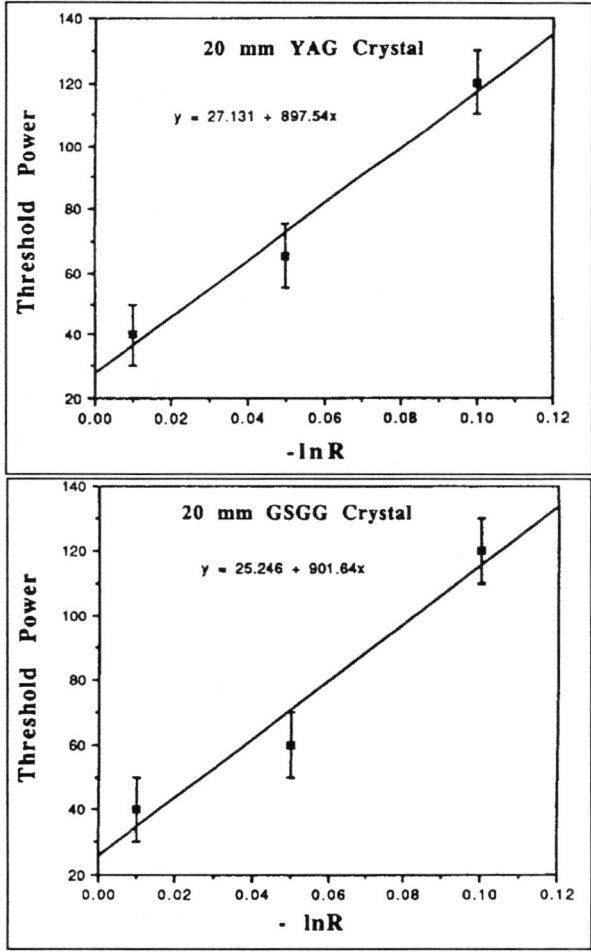

Figure 8. Plots of threshold power versus -lnR. For high values of reflectivity, R, -lnR is approximately equal to the transmission, T, of the output coupler.

Graphs of the threshold power (P_{th}) as a function of the negative of the natural logarithm of the reflectivity of the output coupler (-lnR) are shown in Figure 8 for the 20 millimeter YAG and GSGG crystals. The intercept of each line with the (-lnR) axis gives the roundtrip losses of the laser cavity, 2L; in both cases, 2L ~ .03. Once the roundtrip losses have been determined, the pumping efficiency ε, which is the fraction of absorbed photons averaged over all wavelengths that produce excited atoms participating in the lasing process, can be calculated. It is given by

$$\varepsilon = IA(2L - lnR)/2P_{th}\eta_a\eta_q\eta_{ovp}, \tag{5}$$

where $I = hv/\sigma\tau$, (the fluorescence energy of the crystal divided by the product of the absorption cross-section, σ, and the fluorescence lifetime, τ), A is the cross-sectional area of the crystal, η_a is the absorption efficiency of the crystal, η_q is the quantum efficiency (the mean wavelength of absorbed radiation divided by the lasing wavelength) and η_{ovp} is the overlap of laser absorption with the solar spectrum. [9] Typical values for the constants are:

	Nd:YAG	Nd:Cr:GSGG
I	12.5 W/mm^2	18.0 W/mm^2
η_a	0.59	0.71
η_q	0.63	0.58
η_{ovp}	0.14	0.50

Substitution of these values into equation (5) yields an ε of 0.72 and 0.28 for Nd:YAG and Nd:Cr:GSGG, respectively (for R = 0.95).

Once the pumping efficiency has been determined, the slope efficiency η_s can be calculated by

$$\eta_s = \eta_a\eta_q\eta_{ovp}\varepsilon T/(2L - lnR), \tag{6}$$

where T is the transmission of the output coupler. For Nd:YAG, η_s is about 2 percent, and for Nd:Cr:GSGG, η_s is approximately 3 percent, in reasonably good agreement with the experimental numbers.

The experimental efficiencies exceed those attained in approaches not involving nonimaging optics, with the exception of an unusually high reported set of numbers [10] (Table IV). For instance, one side-pumping approach with Nd:YAG yields an efficiency that is much smaller than 1 percent [11]. Moreover, the efficiencies reported here are comparable to those attained in alternative approaches that do involve nonimaging optics. By employing a nonimaging concentrator in a side-pumping configuration, Weksler and his colleagues at the Weizmann Institute of Science (WIS) report a 1 percent efficiency for sun-pumping Nd:YAG [9] and an efficiency somewhat greater than 2 percent for Nd:Cr:GSGG. The scale of the WIS experiment, however, is vastly larger--a result of the fact that to efficiently absorb light in a side-pumping approach the diameter of the crystal must be comparable to its absorption length, which in this case is about a centimeter. A crystal having such a large diameter requires a huge amount of power to reach the minimum power density required to exceed threshold. In other words, the end-pumping approach offers the efficiency of a large-scale side-pumping approach, but with a table-top experimental apparatus. Indeed, the *overall* efficiency of the end-pumped system actually exceeds that of the WIS system; for every square meter of heliostat mirror the University of Chicago system is able to generate 3 watts of laser power, while the WIS system generates 1 watt per square meter of mirror.

Table IV

Sun-pumped Laser Efficiencies

Approaches without Nonimaging Optics

Investigator (Year)	Crystal	Output P_0 (W)	Collecting Area A (m²)	Input P_i (W)	P_0/A (W/m²)	P_0/P_i (%)
Young[1] (1966)	Nd:YAG	0.8	0.29	140[a]	2.8	0.57
Kozolov[2] (1966)	Nd:CaWO$_4$	0.13	1.77	850[a]	0.07	0.02
Falk[10] (1975)	Nd:YAG	4.5	0.29	140	15.5	3.2
	Nd:Cr:YAG	5	0.29	140	17.2	3.6
Arashi[11] (1984)	Nd:YAG	18	78.5	33,000[b]	0.3	0.05

Approaches with Nonimaging Optics

Investigator (Year)	Crystal	Output P_0 (W)	Collecting Area A (m²)	Input P_i (W)	P_0/A (W/m²)	P_0/P_i (%)
Weksler[9] (1988)	Nd:YAG	60	38.5	5,000	1.6	1.2
	Nd:Cr:GSGG	100	38.5	5,000	2.6	2.0
Cooke (1991)	Nd:YAG	1.7	0.41	200	4.1	0.85
	Nd:Cr:GSGG	3.2	0.41	200	7.8	1.6

----

[a] Since the author does not state P_i, it is taken to be the product of three factors: an incident intensity of 800 W/m², the area of the collecting mirror and an optical relay efficiency of 0.6.

[b] Since the author does not state P_i, it is taken to be the product of the power delivered by the furnace, 55 kW, and an optical relay efficiency of 0.6.

A further improvement in output power is anticipated by utilizing sapphire nonimaging concentrators. Early work in this area was hampered by the lack of a stable optical couplant having an index of refraction of 1.8. A solution of tin chloride dissolved in glycerol was used, but it blackened quickly, causing the concentrator and laser crystal to fracture. It is imperative to develop a good optical couplant having an index of refraction of about 1.8. The couplant must be clear and stable, and it must not burn.

ACKNOWLEDGMENT

I would like to thank several people who made important contributions to this research. Roland Winston, who invented and systematized the study of nonimaging optics, has served as a worthy mentor. Phil Gleckman initiated the solar laser work at the University of Chicago, and Joe O'Gallagher suggested the particular set of experiments I describe here. I am grateful for all the help from the University of Chicago Central Shop; in particular, I wish to thank Helmut Krebs. I am also indebted to Don Levy, Joe Alfano and Niels van Dantzig for graciously allowing me to test the laser crystals with their laboratory argon-ion laser.

I would also like to thank the Link Foundation and the US Department of Energy, especially the National Renewable Energy Laboratory and the Office of Basic Energy Sciences, for support of this work.

REFERENCES

[1]   C. G. Young, "A Sun-Pumped cw 1-W Laser," Appl. Opt. **5**, 993 - 997 (1966).

[2]   N. A. Kozlov, A. A. Mak, B. M. Sedov, "Solid-State, Sun-Pumped Lasers," Sov. J. Opt. Techn. **33**, 549 - 553 (1966).

[3]   R. Winston, "Light Collection within the Framework of Geometrical Optics," J. Opt. Soc. Am. **60**, 245 - 247 (1970).

[4]   A. Rabl, "Comparison of Solar Concentrators," Sol. Energy **18**, 93 - 111 (1976).

[5]   W. T. Welford and R. Winston, *High Collection Nonimaging Optics* (Academic, New York, 1989).

[6]   P. Gleckman, "Achievement of Ultrahigh Solar Concentration with Potential for Efficient Laser Pumping," Appl. Opt. **27**, 4385 - 4391 (1988).

[7]   D. Cooke, and P. Gleckman, H. Krebs, J. O'Gallagher, D. Sagie and R. Winston, "Brighter than the Sun," Nature **346**, 802 (1990).

[8]   R. Winston, "Dielectric Compound Parabolic Concentrators," Appl. Opt. **15**, 291 - 292 (1976).

[9]   M. Weksler and J. Shwartz, "Solar Pumped Solid State Lasers," IEEE J. Quantum Elec. **24**, 1222 - 1228 (1988).

[10]   J. Falk, L. Huff and J. D. Taynai, "Solar-pumped, mode-locked, frequency doubled Nd:YAG laser," in Digest of Technical Papers (Conf. on Laser Eng. and Appl., Washington, DC, 1975), p. 14-15.

[11]   H. Arashi, Y. Oka, N. Sasahara, A. Kaimai and M. Ishigame, "A Solar-Pumped cw 18 W Nd:YAG Laser," Jpn. J. Appl. Phys. **23**, 1051 - 1053 (1984).

A Mathematical Model of a Direct AC-AC Converter Using a Modified Class E Circuit

KELLY J. HERMAN

Department of Electrical Engineering
Michigan Technological University
Houghton MI 49931-1295

ABSTRACT

Present direct ac-ac conversion techniques can suffer from excessive switching losses, power factor problems, high harmonic content, and circuit complexity resulting from the use of multiple switches. We have simulated in our laboratory a modified class E radio-frequency amplifier that achieves direct single-phase ac-ac power conversion at high efficiencies with a single switch. This paper develops an accurate mathematical model that describes the output voltage, input power and output power of a direct ac-ac converter that uses this modified class E topology.

INTRODUCTION

Recent decades have seen definite changes in the concepts and methods used to design and implement power conversion and conditioning equipment. Advances in power semiconductor device technology, in the development of large power integrated circuits, and in resonant and quasiresonant switching techniques, have enabled switching power converters to become quite practical at increased switching frequencies. High-frequency operation provides, at the expense of increased switching losses, the significant advantage of considerably smaller and more efficient passive circuit elements, particularly inductors, in the switching circuits and in the associated filtering networks. The marriage of radio-frequency circuit design principles to conventional power electronic techniques has resulted in a varied and refined collection of dc-dc and dc-ac conversion structures.

Because of the early attention given to dc-dc and dc-ac conversion, principles of ac-ac conversion are not as fully developed. Traditional approaches to electronic ac-ac conversion employ dual conversion structures, which generally consist of an ac-dc converter (rectifier), a dc filter , a dc-ac converter (inverter) and, in some instances, ac filters at input and/or output. Considerable development has gone into dual conversion structures, which have also benefitted from the increased switching speed and power capability of present semiconductor devices. Consequently, this scheme has been, and will continue to be, wholly adequate for many applications in the power electronics industry. Dual conversion structures are not without problems, however. Two converters require additional switching devices, resulting in increased switching losses and decreased conversion efficiency, even in systems employing high-efficiency resonant waveshaping techniques. The dc-link filter further increases losses and also adds considerable size and weight to the system.

In an effort to overcome these drawbacks, various methods of direct ac-ac conversion have been employed. Perhaps the most well-known of these is the ac-ac matrix converter, also known as a nine-switch direct pulse-width-modulated (PWM) converter [1-8]. The ac-ac matrix converter is optimal among three-phase converters from the standpoint of minimum number of switches and minimum filtering requirements. It also benefits from a readily-adjusted input power factor over a range determined by the output power factor. Unfortunately, however, this conversion structure is not readily adaptable to direct single-phase conversion.

There are direct single-phase ac-ac conversion structures which have been described in the literature [8-14]. Generally, however, these converters suffer from increased harmonic levels, have PWM schemes that are difficult to realize, or are restricted in application because the single-phase output can only be obtained at a higher frequency than the three-phase input. In addition, they are still somewhat complex in light of their single-phase output.

Truly single-phase ac-ac conversion falls primarily in one of two categories: ac choppers [15-19] or ac-ac series resonant converters [20-23]. One
drawback to the former approach is the switching losses, which increase
with increasing switching frequency and are compounded by the use of
multiple switches. Some versions also exhibit power factor problems and
contain relatively high harmonic amplitudes due to switching at the input
frequency. The latter converters, when used with thyristors, can provide natural commutation, which reduces circuit complexity. However, this technique generally requires multiple switches.

We have simulated in our laboratory a modified class E radio-frequency
amplifier that achieves direct single-phase ac-ac power conversion at high
efficiencies with a single switch. The input sinusoid can be converted in amplitude and/or frequency. This conversion structure offers advantages over
dual conversion (ac-dc-ac) and other direct conversion schemes by reducing
the number of active devices per phase, and by incorporating the high-efficiency properties characteristic of traditional class E amplifiers. Switch voltage waveforms are shaped by a series-resonant circuit and can be made to
be *zero* at, at least, one of the switch transitions. The converter structure is
evolved from the conventional class E dc-ac inverter, which has been shown
to achieve very high operating efficiencies of 85 to 95 percent at frequencies
of several MHz [24-28].

Before fully applying this discovery to improve the performance of electronic energy processing technology, the development of a useful mathematical model that accurately describes direct ac-ac conversion using a
modified class E topology is required. Such a model is developed in the following sections.

CIRCUIT DESCRIPTION AND PARAMETERS

The class E ac-ac converter, shown in Fig. 1, is driven by an ac source
(V_{ac}) and consists of an active device operated as a switch (S), an inductance in the ac-feed line (L_m), and a load circuit comprising a shunt capacitance (C) (which includes the parasitic output capacitance of the device)
and a series output network $(L_0\text{-}C_0\text{-}R)$. Circuit operation is determined by
the active device switch when "on" and by the transient response of the load
circuit when the switch is "off". Class E operation is achieved when the
switch voltage reaches zero with zero slope at either switch transition,
which minimizes the switching loss even if the switching time is not a small
fraction of the switching period.

The development of the following section uses the idealized equivalent
of the converter shown in Fig. 2. Here, X_m and B are the reactance of L_m
and the susceptance of C, respectively, at the switching frequency. The series output network is replaced by an ideal resonant circuit (resonant at the

switching frequency) and a residual reactance, X. All circuits elements, including the active device, are considered to be ideal.

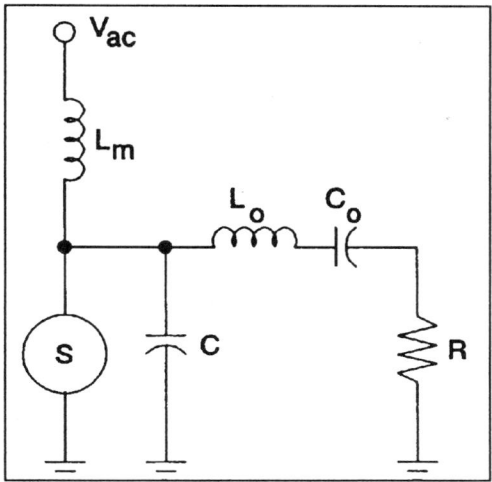

Figure 1.   Direct ac-ac converter using modified class E circuit.

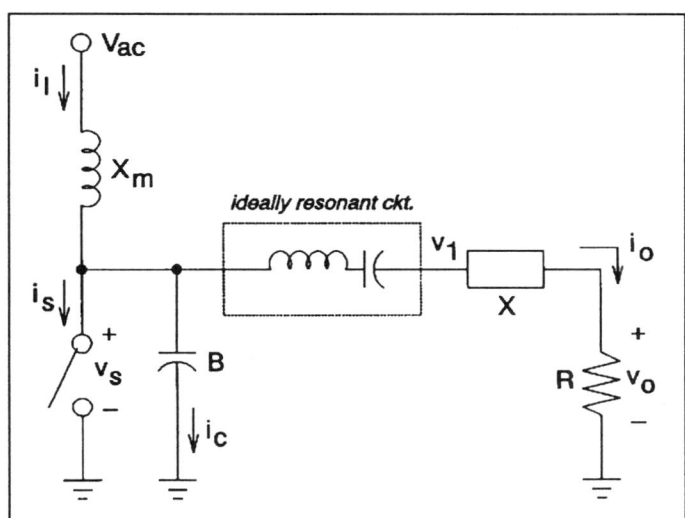

Figure 2.   Idealized equivalent circuit of direct ac-ac converter.

THEORETICAL ANALYSIS

The switching period (Fig. 3) is defined as $y-2\pi \leq \theta \leq y$, where $\theta = \omega t$ represents angular time. The switch is open ('off') in the interval $-y \leq \theta \leq y$, and closed ('on') in the interval $y-2\pi \leq \theta \leq -y$. Thus the switch duty ratio D = y/π. Supply and output voltages are defined as

$$v_s = V_m \sin(n\theta + \phi_s), \tag{1}$$

and

$$v_o = c\sin(\theta + \phi), \tag{2}$$

where n is an arbitrary integer, V_m and c are arbitrary amplitudes, and ϕ_s and ϕ are arbitrary phase shifts relative the midpoint of the switch open interval (i.e., $\theta = 0$), respectively.

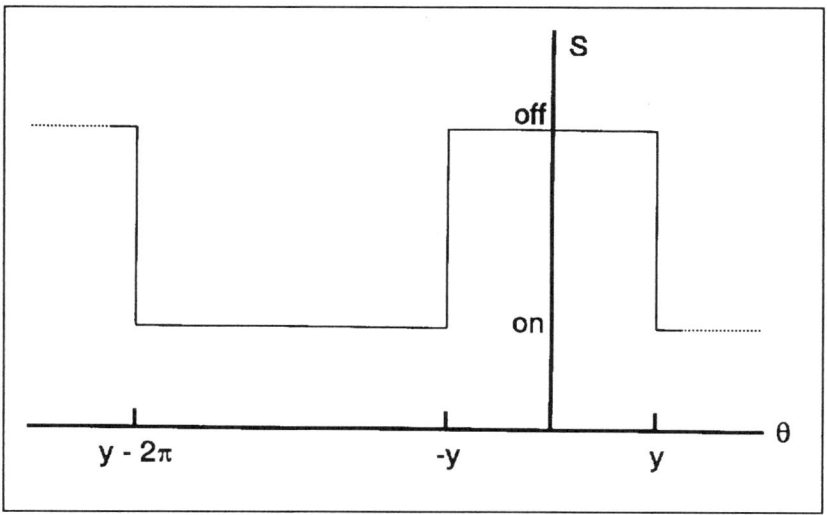

Figure 3.   Switch function in direct ac-ac converter.

During the switch open interval

$$i_{lo} = i_c + i_o, \tag{3}$$

from which

$$\frac{1}{X_m}\int_{-y}^{\theta} [V_m \sin(n\mu + \phi_s) - v_s]\, d\mu + i_{lo}(-y) = Bv_s' + \frac{c}{R}\sin(\theta + \phi), \tag{4}$$

where i_{lo} represents the inductor current i_l during the *open* interval. Similarly, during the switch closed interval

$$i_{lc} = \frac{1}{X_m} \int_{y-2\pi}^{\theta} V_m \sin(n\mu + \phi_s)\, d\mu + i_{lc}(y - 2\pi), \qquad (5)$$

where i_{lc} represents i_l during the *closed* interval. Because neither the current through X_m nor the voltage across B can change instantaneously, we have the boundary conditions

$$i_{lc}(-y) = i_{lo}(-y), \qquad (6)$$

$$i_{lc}(y - 2\pi) = i_{lc}(y) = i_{lo}(y), \qquad (7)$$

and

$$v_s(y) = 0. \qquad (8)$$

Combining [4-8] gives a second-order differential equation, which may be solved for the switch voltage v_s using standard techniques (such as Laplace transforms):

$$v_s = \frac{b_1 \cos(k\theta) + b_2 \sin(k\theta) - k^2 V_m \sin(n\theta + \phi_s)}{n^2 + k^2} + \frac{c\cos(\theta + \phi)}{BR(1 - k^2)}, \qquad (9)$$

where

$$b_1 = \frac{k^3 V_m}{n(n^2 - k^2)} \frac{\sin(ny)\sin(\phi_s)}{\sin(ky)} - \frac{ck}{BR(1 - k^2)} \frac{\sin(y)\cos(\phi)}{\sin(ky)}, \qquad (10)$$

$$b_2 = \frac{V_m}{n^2 - k^2} \frac{k^2 \sin(ny - \phi_s) + k^3 \cos(ky)\sin(ny)\sin(\phi_s)}{n\sin(ky)}$$

$$+ \frac{c}{BR(1 - k^2)} \frac{\cos(y - \phi) - [k\cos(ky)\cos(\phi)] / \sin(ky)}{\sin(ky)}, \qquad (11)$$

and

$$k = \frac{1}{\sqrt{BX_m}}, \qquad (12)$$

are used for convenience.

Because the output voltage is a sinusoid, theoretical voltage v_1 is also a sinusoid:

$$v_1 = c_1 \sin (\theta + \phi_1), \tag{13}$$

where

$$c_1 = \rho c, \tag{14}$$

$$\rho = \sqrt{1 + \frac{X^2}{R^2}}, \tag{15}$$

$$\phi_1 = \phi + \psi, \tag{16}$$

and

$$\psi = \tan^{-1} (\frac{X}{R}) \tag{17}$$

Because the ideal resonant circuit presents zero impedance to fundamental frequency current, there can be no fundamental frequency voltage drop across it. Thus the fundamental frequency component of v_s must be the theoretical voltage v_1. The magnitude of this voltage can be determined by the Fourier integral

$$c_1 = \rho c = \frac{1}{\pi} \int_0^{2\pi} v_s(\theta) \sin (\theta + \phi_1) \, d\theta. \tag{18}$$

Further, because the fundamental frequency component of v_s is, by definition, a sinusoid of phase ϕ_1, there can be no quadrature component with respect to ϕ_1. Thus, a second Fourier integral can be written:

$$0 = \frac{1}{\pi} \int_0^{2\pi} v_s(\theta) \cos (\theta + \phi_1) \, d\theta \tag{19}$$

Once the switch duty ratio, ac supply voltage, and component values have been specified, (18) and (19) represent two equations in the two unknowns, c and ϕ. These equations can be solved analytically using the techniques shown in [27-31]. Because of their length and complexity, equations for c and ϕ are presented in the Appendix.

With output voltage fully specified by c and ϕ, only powers remain to be determined. Equations for input and output powers are straightforward:

$$P_{in} = \frac{1}{2\pi} \int_0^{2\pi} V_m \sin (n\theta + \phi_s) i_I(\theta) \, d\theta, \tag{20}$$

and

$$P_{out} = \frac{c^2}{2R}.$$ (21)

Thus, the circuit efficiency is

$$\eta = \frac{P_{out}}{P_{in}}.$$ (22)

CONCLUSIONS

An analytical model has been developed for a direct, single-phase ac-ac converter using a modified class E radio frequency amplifier. Specification of the circuit parameters and supply voltage allows output voltage, input and output power, and circuit efficiency to be determined. The analytical model is only a part of the problem, however. Before fully applying this conversion technique to electronic energy processing systems, it is also necessary to: 1) develop appropriate techniques to optimize the performance of these circuits; 2) perform relevant experimental studies related to the mathematical models; and 3) evaluate the relative contributions and potential improvements of this approach to direct, single-phase ac-ac conversion. The author's colleagues are continuing work in these areas. Subsequent papers are planned.

ACKNOWLEDGEMENTS

To Robert E. Zulinski, who was my advisor, for teaching me class E amplifiers; to Michael, for opening my eyes to who I am and where I want to go; and to those who supported me through the years to get me where I am - there are far too many to thank each one individually.

APPENDIX

The Fourier integrals [18] and [19] are solved for $\emptyset$ by writing them in the form

$$a_1 \cos(\phi) + a_2 \sin(\phi) = 0,$$ (A1)

from which

$$\phi = \tan^{-1}(a_1/a_2), \tag{A2}$$

where

$$a_1 = s_{1c}(s_6-s_8)-s_{2c}(s_5+s_7) + (1-k^2)(s_{4c}+y)(s_5-s_7)\sin(ky)$$

$$+ (1-k^2)^2 BR\rho \sin(ky)[(s_7-s_5)\sin(\psi) + (s_6-s_8)\cos(\psi)], \tag{A3}$$

$$a_2 = s_{1s}(s_8-s_6) - (1-k^2)(s_{3s}-y)(s_8-s_6)$$

$$-(1-k^2)^2 BR\rho \sin(ky)[(s_7-s_5)\cos(\psi) + (s_8-s_6)\sin(\psi)], \tag{A4}$$

$$s_{1c} = \frac{2k\cos(ky)\sin(y)[k\sin(y)\cos(ky) - \cos(y)\sin(ky)]}{\sin(ky) - \cos(y)}, \tag{A5}$$

$$s_{1s} = 2\sin(y)[\cos(y)\sin(ky) - k\sin(y)\cos(ky)], \tag{A6}$$

$$s_{2c} = 2k\sin(y)[\sin(y)\cos(ky) - k\cos(y)\sin(ky)], \tag{A7}$$

$$s_{3s} = s_{4c} = \sin(y)\cos(y), \tag{A8}$$

$$s_5 = \left[\sin(ny-\phi_s) + \frac{k\cos(ky)\sin(ny)\sin(\phi_s)}{n\sin(ky)}\right]$$

$$\times [k\sin(y)\cos(ky) - \cos(y)\sin(ky)] \tag{A9}$$

$$s_6 = (k/n)\sin(ny)\sin(\phi_s)[\sin(y)\cos(ky) - k\cos(y)\sin(ky)], \tag{A10}$$

$$s_7 = \frac{(1-k^2)\sin(ky)\cos(\phi_s)[n\sin(y)\cos(ny) - \cos(y)\sin(ny)]}{1-n^2}, \tag{A11}$$

and

$$s_8 = \frac{(1-k^2)\sin(ky)\sin(\phi_s)[\sin(y)\cos(ny) - n\cos(y)\sin(ny)]}{1-n^2}. \tag{A12}$$

With a solution for ⌀, c is determined by:

$$c = \frac{2p_4 V_m BRk^2 (1 - k^2)}{p_3 (n^2 - k^2)}, \tag{A13}$$

where

$$p_3 = -[s_{1c}\cos(\phi) + s_{1s}\sin(\phi)]\sin(\phi_1) + s_{2c}\cos(\phi)\cos(\phi_1)$$

$$, \tag{A14}$$

$$-(1 - k^2)\sin(ky)[y\cos(\psi) - s_{3s}\sin(\phi)\sin(\phi_1) + s_{4c}\cos(\phi)\cos(\phi_1)]$$

and

$$p_4 = (s_7 - s_5)\sin(\phi_1) + (s_6 - s_8)\cos(\phi_1). \tag{A15}$$

REFERENCES

[1]   L. Gyugyi and B. Pelly, *Static Power Frequency Changers*, Wiley, New York, 1976.

[2]   M. Venturini, "Convertitore diretto ac-ac di elevata potenza," Italian Patent 20777a-79, March 6, 1979.

[3]   M. Venturini, "A new sine wave in, sine wave out, conversion technique eliminates reactive elements," in *Proc. Powercon 7*, pp. E3-1-E3-15, (San Diego, CA), 1980.

[4]   A. Alesina and M. Venturini, "Solid-state power conversion: a Fourier analysis approach to generalized transformer synthesis," *IEEE Trans. Circuits and Systems*, vol. CAS-28, pp. 319-330, April 1981.

[5]   M. Venturini and A. Alesina, "Method and apparatus for the conversion of a polyphase voltage system," U.S. Patent 4,628,425, Dec. 9, 1986.

[6]   A. Alesina and M. Venturini, "Intrinsic amplitude limits and optimum design of 9-switches direct PWM ac-ac converters," *IEEE Power Electronics Specialists Conf. Record*, vol. 2, pp. 1284-1291, (Kyoto, Japan), April 1988.

[7]   A. Alesina and M. Venturini, "Analysis and design of optimum-amplitude nine-switch direct ac-ac converters," *IEEE Trans. Power Electronics*, vol. 4, pp.101-112, Jan. 1989.

[8]   P.D. Ziogas, S.I. Khan, and M.H. Rashid, "Some improved force-commutated cycloconverter structures," *IEEE Trans. Industry Applications*, vol. IA-21, pp. 1242-1253, Sept./Oct. 1985.

[9]   P.D. Ziogas, S.I. Khan, and M.H. Rashid, "Analysis and design of forced commutated cycloconverter structures with improved transfer characteristics," *IEEE Power Electronics Specialists Conf. Record*, pp. 610-621, (Toulouse, France), June 1985.

[10]   A. Ishiguro *et al.*, "A new method of PWM control for forced commutated cycloconverters using microprocessors," *IEEE IAS Annual Meeting Conference Record*, pp. 520-526, (Pittsburgh, PA), Oct. 1988.

[11]   T. Tanaka and H. Kasahara, "An ac to ac converter improved in input current harmonics and power factor," *IEEE Power Electronics Specialists Conf. Record*, pp. 212-218, (San Diego, CA), June 1979.

[12]   S.B. Dewan and J.E. Quaicoe, "A six-pulse load-commutated cycloinverter with a resistive load," International Power Electronics Conference, (Tokyo, Japan), March 1983.

[13]   V.S. Gharpure, B. Tech, and G.N. Revankar, "A cycloconverter-linked induction heating system," *IEE Proceedings*, vol. 130, pp. 321-330, Sept. 1983.

[14]   S. Okada, T. Tanaka, and H. Kasahara, "A new high frequency cycloconverter," *IEEE Trans. Power Electronics*, vol. PE-1, pp. 215-222, Oct. 1986.

[15]   D.W. Borst, E.J. Diebold, and F.W. Parrish, "Voltage control by means of power thyristors," *IEEE Trans. Industry and General Applications*, vol. IGA-2, pp. 102-124, Mar./Apr. 1966.

[16]   G.N. Revankar and D.S. Trasi, "Symmetrically pulse width modulated ac chopper," *IEEE Trans. Industrial Electronics and Control Instrumentation*, vol. IECI-24, pp. 39-45, Feb. 1977.

[17]   K.A. Krishnamurthy, G.K. Dubey, and G.N. Revankar, "Ac power control of an R-L load," *IEEE Trans. Industrial Electronics and Control Instrumentation*, vol. IECI-24, pp. 138-141, Feb. 1977.

[18]   G.H. Choe *et al.*, "A new pulse-width modulated method for ac chopper," *IEEE Power Electronics Specialists Conf. Record*, pp. 156-164, (Blacksburg, VA), June 1987.

[19]   A. Khoei and S. Yuvarajan, "Single-phase ac-ac converters using power MOSFET's," *IEEE Trans. Industrial Electronics*, vol. 35, pp. 442-443, Aug. 1988.

[20]   J.B. Klaassens, Dc-to-ac series-resonant converter system with high internal frequency generating synthesized waveforms for multikilowatt power levels," *IEEE Trans. Power Electronics*, vol. PE-1, pp. 9-20, Jan. 1986.

[21]   J.B. Klaassens and E.J.F.M. Smits, "Series-resonant ac-power interface with an optimal power factor and enhanced conversion ratio," *IEEE Power Electronics Specialists Conf. Record*, pp. 39-48, (Vancouver, BC), June 1986.

[22] J.B. Klaassens, "Dc-ac series-resonant converter system with high internal frequency generating multiphase ac-waveforms for multikilowatt power levels," *IEEE Trans. Power Electronics*, vol. PE-2, pp. 247-256, July 1987.

[23] J.B. Klaassens and J. van Duivenbode, "Series-resonant energy conversion with multi-segment current waveforms for bipolar energy flow," *IEEE Power Electronics Specialists Conf. Record*, pp. 599-608, (Kyoto, Japan), April 1988.

[24] N.O. Sokal and A.D. Sokal, "Class E - a new class of high-efficiency tuned single-ended switching power amplifiers," *IEEE Jour. Solid-State Circuits*, vol. SC-10, pp. 168-176, June 1975.

[25] R.J. Gutmann, "Application of rf circuit design principles to distributed power converters," *IEEE Trans. Industrial Electronics and Control Instrumentation*, vol. IECI-27, pp. 156-164, Aug. 1980.

[26] R. Redl, B. Molnar, and N.O. Sokal, "Class E resonant regulated dc/dc power converters: analysis of operation, and experimental results at 1.5 MHz," *IEEE Power Electronics Specialists Conf. Record*, pp. 50-60, (Albuquerque, NM), June 1983.

[27] R.E. Zulinski and J.W. Steadman, "Idealized operation of class E frequency multipliers," *IEEE Trans. Circuits and Systems*, vol. CAS-33, pp. 1209-1218, Dec. 1986.

[28] R.E. Zulinski and J.W. Steadman, "Class E power amplifiers and frequency multipliers with finite dc-feed inductance," *IEEE Trans. Circuits and Systems*, vol. CAS-34, pp. 1074-1087, Sept. 1987.

[29] F.H. Raab, "Effects of circuit variations on the class E tuned power amplifier," *IEEE Jour. Solid-State Circuits*, vol. SC-13, pp. 239-247, April 1978.

[30] R.E. Zulinski, "The effects of parameter variations on the performance of diode-modified class E tuned power amplifiers," M.S. Thesis, Michigan Technological Univ., Houghton, Oct. 1980.

[31] R.E. Zulinski, "Class E frequency multipliers," Ph.D. dissertation, University of Wyoming, Laramie, May 1985.

Aluminosilicate Membrane Preparation and Characterization

SUSAN L. HIETALA

Department of Chemical and Nuclear Engineering
University of New Mexico
Research Supervisor: Dr. Douglas M. Smith

ABSTRACT

A sol-gel method for preparing aluminosilicates was investigated for viability of producing microporous membranes. The data suggests that the membranes prepared were in the microporous regime, and continued advances in the novel characterization methods used are ongoing. A surface acoustic wave (SAW) technique, which was utilized to resolve pore size information in thin films, was incorporated into commercial volumetric adsorption instrumentation. This surface sensitive technique can inexpensively elucidate the physical characteristics of membranes (e.g. surface area and pore size), and implement data reduction techniques commonly used for commercial bulk analysis of materials. The aluminosilicate membranes were also characterized using single-gas permeability measurements.

INTRODUCTION

MEMBRANES

Gas and liquid separations are generally energy intensive processes in which conservation and efficiency improving techniques are sorely needed. Developments of the area of membrane separations have been noted, yet membrane techniques could be more widely utilized if there were selectively controlled pore sizes, smaller pore sizes, and more robust materials. Inorganic (particularly ceramic) membranes have these advantages, and when processed via a sol-gel technique, offer a controlled and versatile method for preparation.

With the tremendous increase in membrane use in the past couple decades, anticipated to continue over the next several years, the U.S. Department of Energy commissioned researchers to identify membrane-related areas needing further development.[1] In the report entitled "Membrane Separation Systems: A Research Needs Assessment", a prioritized list has been assimilated, with the top ten as follows:

- Pervaporation membranes for organic-organic separations
- Oxidation-resistant reverse osmosis membrane
- Gas separation membranes with skins <500 Å
- Oxygen selective facilitated transport membranes
- Gas separation with higher oxygen/nitrogen selectivity
- Fouling-resistant ultrafiltration membranes
- Pervaporation modules with better solvent resistance
- General fabrication method for composite hollow fibers with less than 500 Å skins
- Microfiltration at high temperatures
- Low-cost microfiltration modules

In general, a variety of inorganic membranes are being studied to address these issues. In particular, the work described in this paper focuses on preparing ceramic microporous membranes via a sol-gel method. This type of membrane would be especially suited for gas separations/selectivity at high temperatures and pressures.

For separating submicron-size materials, traditionally organic membranes have been used. The ease of processing (low temperatures) and effectiveness (solution/dissolution properties) of organic polymeric membranes are also the same properties that lead to several inherent disadvantages including: (1) narrow thermal stability, (2) limited chemical stability (such as swelling in organic solvents), and (3) limited mechanical stability to compressive forces. These membranes are thus unsuitable where high temperatures or aggressive chemical environments are encountered.

The ineffectiveness of organic polymeric membranes in some applications has led to increased research and development of inorganic (e.g. ceramic) membranes, especially for reverse osmosis, ultrafiltration, microfiltration and conventional filtration. Many industries already extensively incorporate ceramic membranes for separation processes. [1-3] The *food and beverage* industry uses ceramic membranes for clarification and sterilization of fruit juices and vinegars; to concentrate and separate dairy products; remove unwanted phenols, colors, tannins or other impurities in wines. Ceramic membranes are used in the *biotechnology* industry to concentrate and purify vaccines and vitamins; or remove viruses from culture broths. *Gas separation* industries attempt to employ membranes to remove hydrogen from refinery streams; carbon dioxide and hydrogen sulfide from natural gas; and for nitrogen enrichment or methane recovery in mining operations. (This industry, in particular, could benefit from improved microporous membranes). Other industries that prominently use ceramic membranes include *environmental control, petrochemical, metal refining and electronic.*

Membranes separations are attained by a variety of mechanisms, the primary factors are size, diffusivity, ionic change, vapor temperature and pressure, solubility, surface activity and density. The effect of these factors on separation efficiency are shown in Figure 1, a modified version of a table originally published in 1969 by Dorr-Oliver, Inc. Microporosity, as discussed in this paper, refers to the 'ultrafiltration' regime or a pore size of less than 20 Å.

With the benefits of greater thermal, mechanical and chemical stability and possibility of introducing catalytic electrochemical sites, ceramic membranes do have disadvantages. In general, ceramics are brittle and expensive to produce and repair. However, with continued research and development, the material properties will be improved and the costs will be reduced. Some commonly used membrane materials include silica, alumina, zirconia, and other glass formers. This research focuses on the sol-gel prepared aluminosilicates, which have demonstrated some unusual properties in the bulk material (including microporosity), thus an interesting system for membrane-related studies. Some of the useful properties in this system (discussed later) include the acid/base characteristics [4], the myriad of synthesis schemes and precursors [5], chemical and thermal stability [6], and the compositional interest.[7]

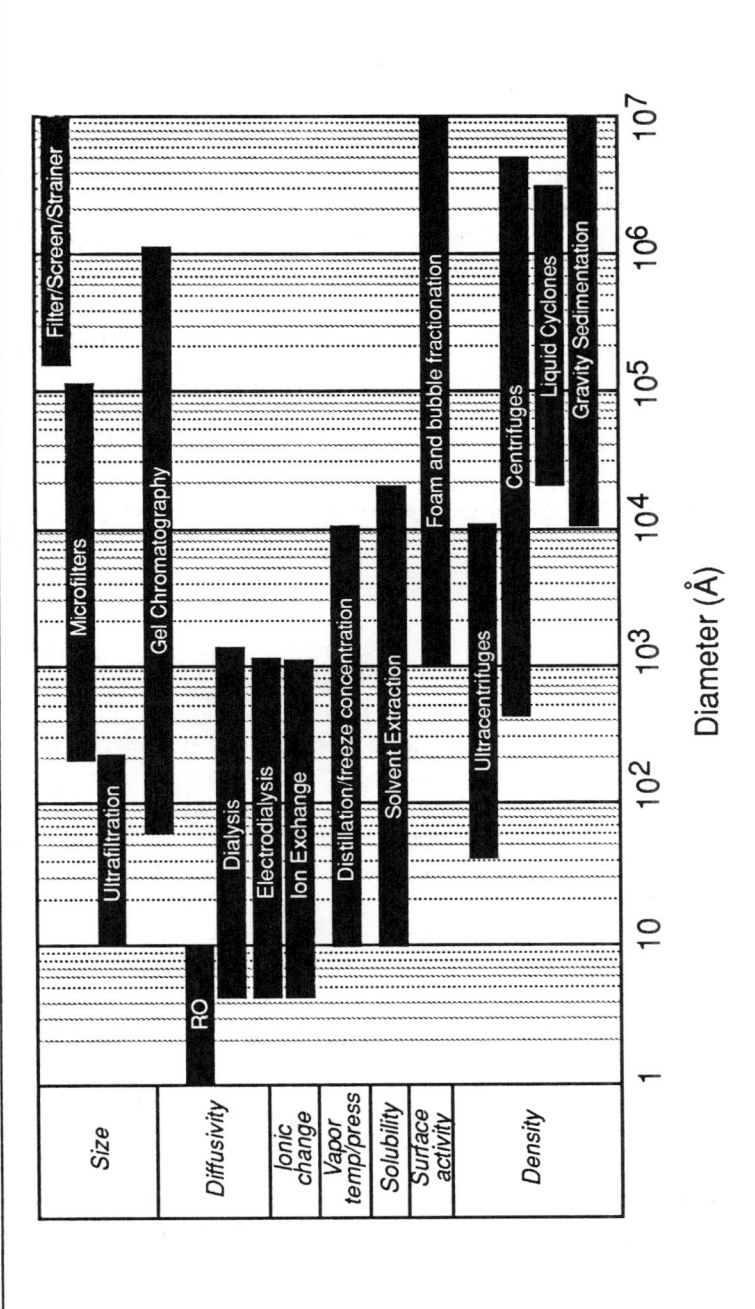

Figure 1. Primary factors affecting separations, and their relative size dependence (after Dorr-Oliver, 1969).

CHARACTERIZATION

The characterization of membranes is necessary to determine and effectively modify size exclusion and chemical properties. Currently membrane parameters are not well standardized, and a trial and error approach is typically used to select appropriate properties. A panel of specialist in the membrane area brought together to discuss the characterization of membranes, emphasized that [8]

> *"Pore size distribution curves are badly needed for microfiltration membranes..."*

The characterization of pore structure in thin films (<1 μm) is difficult, and in many cases the information is ambiguous. Often, bulk characterization techniques have been inappropriately applied to thin films, without regard to the small total pore volume and surface area contained in the pores. For example, commercial gas adsorption/desorption and mercury porosimetry instruments commonly have resolution limits $\approx 10^{-4}$ m^2/g. In order to accurately probe a 1 μm film of 10% porosity with this technique, a film area >100 cm^2 would be required (and of course much larger areas would be required for *thin* films). Such a large size is not readily handled by most commercial volumetric and gravimetric instruments. A novel technique which has been used to probe the structure of films is acoustic waves. Both bulk acoustic waves (commonly in the form of a quartz crystal microbalance)[9] and surface acoustic waves (SAW)[10] have been used. In these techniques nitrogen sorption at 77K is typically used to study film texture or pore size/surface area. While the technique has the necessary resolution, the implementation has suffered from certain drawbacks primarily due to its limited pressure ranges and lack of commercial instrumentation. To overcome these shortcomings, a SAW system has been designed and incorporated into a Micromeritics ASAP 2000, which has the capabilities of probing nonporous, microporous and mesoporous materials. The advantages of this system are that it incorporates:

1. measurement of adsorption over a very wide relative pressure range with the ability to accurately measure uptake in the 10^{-6} to 10^{-1} P/P_0 range,
2. measurement of relative pressures very close to 1.0 to assess possible mesoporosity,
3. ability to use a large variety of gases and temperatures,
4. and access to the large base of commercially available software for data reduction of isotherms to yield pore structure information.

Another important aspect of characterizing membranes is determining their separation capabilities. This was ascertained by a system designed and built by Brinker, Ward, Sehgal and Raman at the University of New Mexico. The permeability and separation factors using various gases were obtained.

The knowledge of factors affecting porosity and film formation, combined with SAW and permeability characterization, allows modifications of pore characteristics and evaluation of these effects on permeability. The emphasis of this research is the characterization of aluminosilicate structures, and the effect of various processing conditions on the bulk and film properties.

SOL-GEL PREPARED ALUMINOSILICATES

A closer look at the properties of aluminosilicates illustrates why they warrant further study for membrane applications.

(1). Acid/Base Properties

The acid/base characteristics of aluminosilicates demonstrate a unique potential as membrane reactors. Aluminum in a tetrahedrally coordinated site in an aluminosilicate network forms a Lewis acid site. The interactions of a molecule adsorbed on this site with other molecules, forms a Brönsted acid site. This inherent property of the aluminosilicate membranes not only selectively bonds molecules, but may also enhance reactivity in some conditions.

(2). Synthesis

The myriad of precursors available for sol-gel synthesis allow tailoring of the synthesis procedure to produce specific reaction conditions, modifying the membrane properties. Properties that may be altered include (but are not limited to): pore size, pore size distribution, reactivity, hardness and fracture toughness.

(3). Stability

The refractory nature of aluminosilicates enable these materials to maintain their stability at high temperatures and pressures, while resisting chemical degradation. These properties promote uses in harsh environments and high pressures where traditional organic membranes are not appropriate.

(4). Compositional Interest

Of special interest is the anomalous compositional behavior of sol-gel prepared aluminosilicates. Figure 2 shows the change in surface area and density from sol-gel prepared aluminosilicate materials through a range of aluminum to silicon ratios. Compositional variation of the material showed anomalies not only in surface area and densities, but physical structure and phase evolution as well.[11] The decrease in surface area and density at the aluminum to silicon molar ratio of 1:1 is especially dramatic, since less than a 5% compositional variation causes a several orders of magnitude change in the surface area. Experiments have shown this effect is due to the closed porosity in the 1:1 material, and not due to crystalline phases.[12]

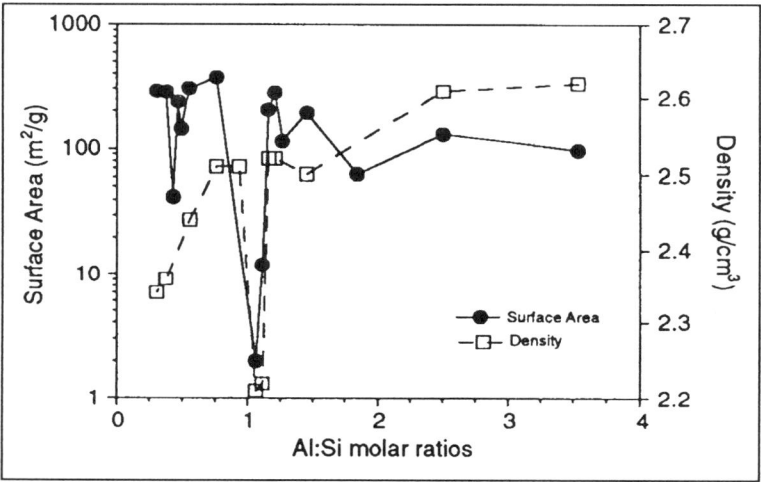

Figure 2. Surface area and density in various aluminosilicate compositions prepared via a sol-gel route.

RESULTS FROM BULK ALUMINOSILICATES

The final properties of sol-gel prepared materials are dependent on a number of variables including: precursor reactivity, solvent reactivity, temperature, pH, drying and curing kinetics. Typical reactions occurring during formation of a sol-gel prepared ceramic are simply illustrated in Figure 3 for a silicate system. Similar types of reactions are expected in an aluminosilicate system, the kinetics of the specific reactions are however quite different. Note that the relative acidity and basicity of the ligands during the sol-gel synthesis are going to have significant effects on not only the type of reactions, but rate of reactions as well (Figure 4). This flexibility with the reactivity of the precursors is one way to tailor the membrane properties.

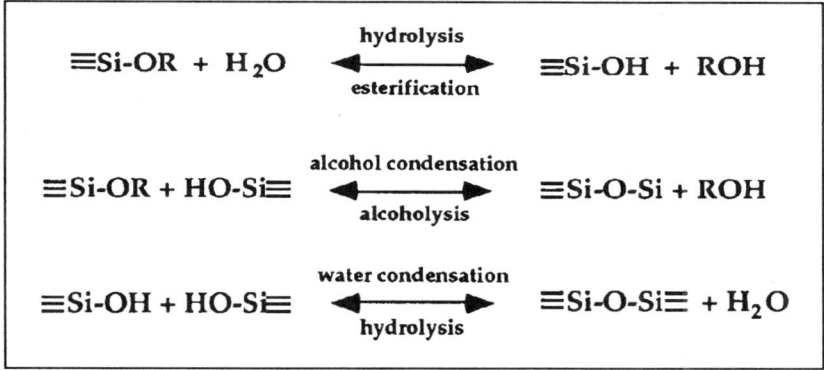

Figure 3. Typical reactions for sol-gel prepared silica gels.

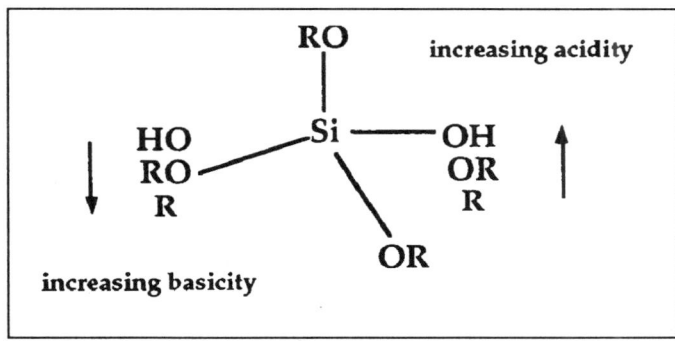

Figure 4. Relative acid/base characteristics of various ligands.

The effects of altering reactions after gel formation (recall Figure 3) can be observed from experiments such as 'washing' the gel. Figure 5 is a plot of nitrogen isotherms for xerogels (dried gels), where the original pore fluid was replaced by water (via washing) during gelation phase. Properties of the 1:1 xerogel composition that resulted from variations in the water to silicon ratio ranged from nonporous (100 H_2O:Si) to microporous (60 H_2O:Si), and with a slight change in processing conditions (e.g. washing with ethanol), to mesoporous. A narrow pore size distribution in the 60 H_2O:Si sample was determined by the similarity of the adsorption and desorption isotherms, illustrated in Figure 6. A narrow pore size distribution is a property especially desirable for membranes, a unique property of this sol-gel prepared aluminosilicate.

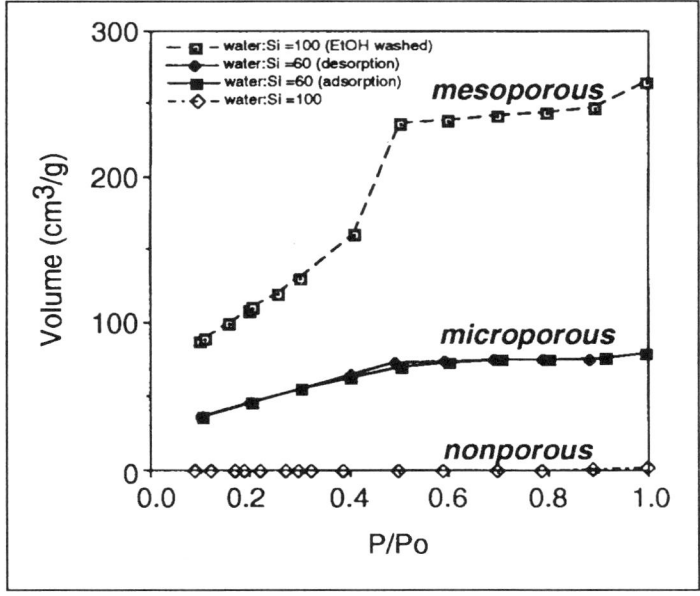

Figure 5. Effect of processing variations (water: silicon ratios and "washing") on the accessible porosity of 1:1 bulk aluminosilicate.

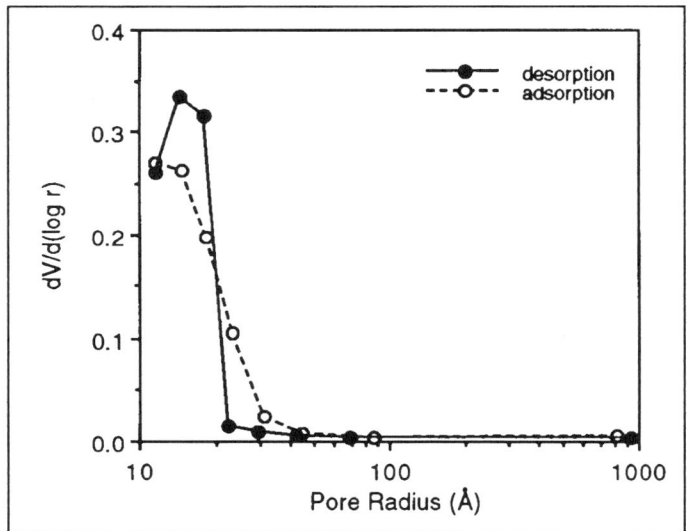

Figure 6. Pore size distribution for 60 water:Si adsorption/desorption isotherms.

In general, the 1:1 aluminosilicate composition with water to silicon molar ratios of less than 100 exhibits open porosity, while ratios greater than 100 exhibit closed porosity. This is illustrated in Figure 7, which shows the accessible pore volume in the materials. The nitrogen uptake (pore volume) of gels prepared with large water to silicon ratios is nearly immeasurable, whereas for gels prepared with lower ratios the accessible porosity becomes more significant, with the most pronounced uptake found in the ethanol washed gel.

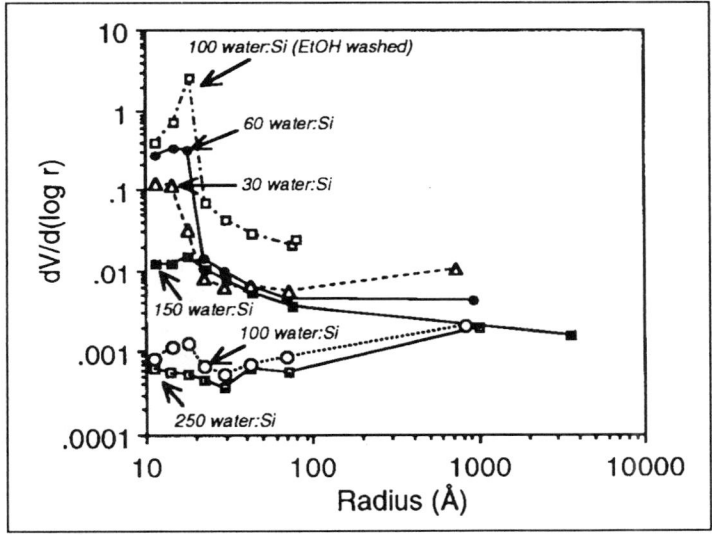

Figure 7.   Volume of gas adsorbed as a function of pore radius for various water to silicon ratios in a 1:1 aluminosilicate.

The drying rate, stiffness of the gel, size of the body, permeability of the structure and the size and distribution of pores or defects all contribute to the final structure of the dried material.[13] The pressure (P) induced in a pore during drying is related to the pore radius (r), the surface tension (γ) and the contact angle (ϕ) of liquid (illustrated in Figure 8) by the equation:

$$P = \frac{2\Delta\gamma}{r}. \tag{1}$$

The results of applying the surface tension/contact angle ideas to modify pores can clearly be seen in the formation of aerogels.[14-16] In this technique the solvent of a gel may be exchanged with one which can be brought to its critical point at reasonable temperatures and pressures (CO_2 is commonly used). The system is brought to the critical point, without crossing a phase boundary, and while at the critical point, the solvent is vented at constant temperature. In this case there is no liquid-vapor interface and no capillary pressure to collapse the pore, thus the pore remains open. Results from this technique have indeed shown large pore structures, high insulative values, and for silica, a refractive index of <1.1 (air is 1.0).

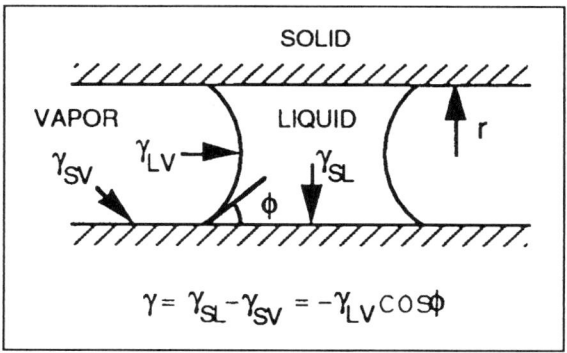

Figure 8. Illustration of the interactions of solid, vapor and liquid in a cylindrical pore (after Brinker, et al, reference 13).

To explore the concept of surface tension in aluminosilicates, the structure of the 1:1 aluminum to silicon composition was modified by varying the pore fluids during drying. Figure 9 shows the results of washing a 1:1 aluminosilicate with fluids of varying surface tensions before drying. The graph illustrates how higher surface tension fluids (such as water) have an effect of closing pores or collapsing a structure during drying, whereas lower surface tension pore fluids tend to collapse the structure less upon drying, thereby exhibiting larger surface areas and pore volumes. The effects, of course, can be not only surface tension (and contact angle) driven, but also reaction driven. Refer back to the reactions discussed in Figure 3, noting how an excess of solvent (such as alcohol) may shift the reactions.

A series of experiments aging a 1:1 aluminosilicate in water solutions of various pH is shown in Figure 10. The pH affects the solubilities of the components (e.g. silica and alumina) in the gel. The gels were either washed with the solutions and dried directly, or they were allowed to remain in contact with an excess of the pore fluid for varying amounts of time prior to drying. The surface areas of the xerogels were $\approx 1 \ m^2/g$, and the porosities were calculated from the measured skeletal density. The isoelectric point (6.0) was measured on an unwashed xerogel in water by an electrokinetic sonic amplitude method.[17] As the conditions increased in basicity, the gels tended towards higher porosity due to dissolution/reprecipitation reactions.

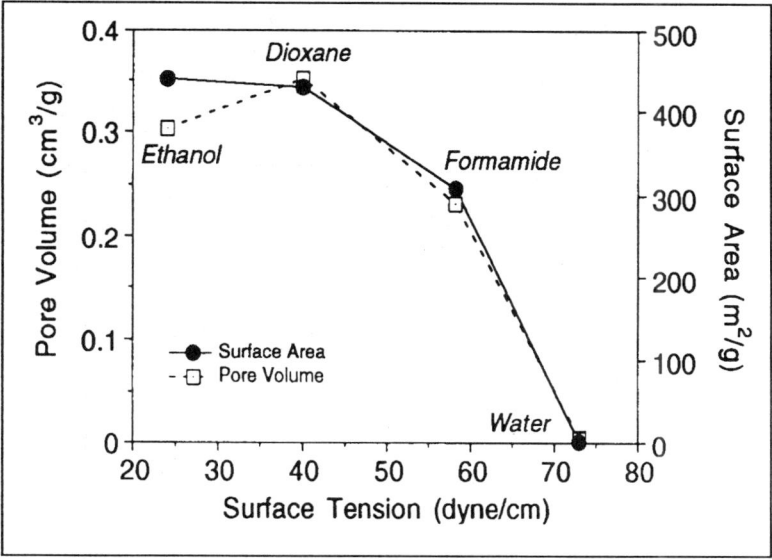

Figure 9.   Effect of surface tension of gel wash fluids on surface area and pore volume of xerogels.

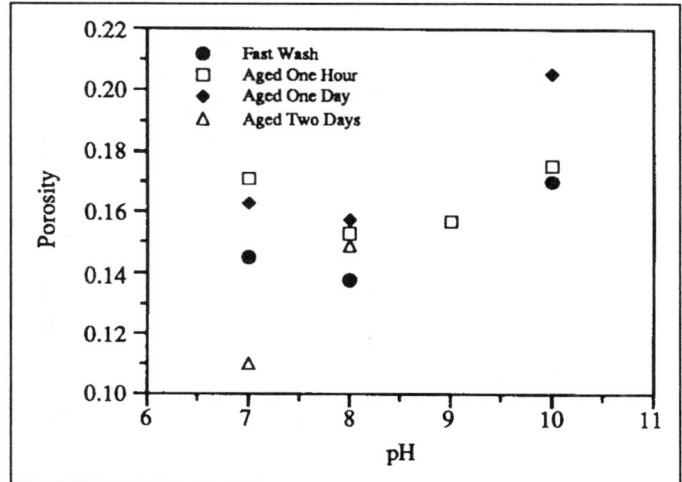

Figure 10.   Calculated porosity for a 1:1 aluminosilicate composition aged in water at various pH values.

From these types of experiments, we find that the properties of sol-gel prepared aluminosilicates may be modified by change in composition, pore fluid, aging and drying conditions. The 1:1 composition can be synthesized to have either open or closed pores, with varying amounts of porosity. Using processing ideas developed from bulk data, we have a basis for film formation with control of critical properties.

CHARACTERIZATION

SAW

A surface acoustic wave (SAW) technique was used to elucidate surface area and pore size information from thin films. Since the use of acoustic waves for determining surface properties of films is relatively novel, a short history and considerations for implementing surface acoustic waves as a viable measurement technique are discussed.

BACKGROUND OF SAW

Surface acoustic waves were originally described by Lord Rayleigh in 1885.[18] Work was progressed in this area, primarily by seismologists, until recently when these waves were found useful for applications in electronics. Since then, SAW devices have been extensively used for filters,[19, 20] delay lines [21, 22] and most recently as sensitive chemical sensors.[23-28]

A surface acoustic wave is described as a wave traveling across a surface, where the atoms and their center of mass move together, as in long wavelength acoustical vibrations. The term "acoustic" follows the common usage in physics, where *ordinary* elastic motions in crystal are called acoustic modes.[29, 30]

The easiest (and most common) mode of surface wave to excite is called the Rayleigh wave. This wave has both particle displacements and electric field components confined to the sagittal plane (the plane containing the surface normal and the propagation direction), and can be described as a "linear combination of bulk acoustic waves, which satisfy the stress free boundary condition".[31] For these waves, >95% of the energy is confined within a depth equal to one acoustic wavelength, which for quartz is 32 mm. The propagation of elastic waves across a surface has been rigorously derived in texts by Auld.[32]

Surface acoustic waves are efficiently generated using piezoelectrics. For a material to be piezoelectric, it must be a dielectric, and in one of the 21 non-centrosymmetric point groups (where 20 of the 21 are piezoelectric). In general, a mechanical stress applied to a piezoelectric crystal causes it to polarize and develop an electrical charge on the opposite crystal face. Conversely, an electrical charge or polarization, creates a mechanical stress in the material. The piezoelectric effect is dependent on the crystal structure and the direction of the stress. For example, quartz develops a polarization when subjected to a compressive stress along [100] plane, but not when stressed along [001]. The piezoelectric effect can be described by the following equations:

$$P = Zd + E\chi, \tag{2}$$

$$e = Zs + Ed, \tag{3}$$

where, P is the polarization, Z is the stress, d is the piezoelectric strain constant, E is the electric field, χ is the dielectric susceptibility, e is the elastic strain, and s is the elastic compliance constant. A general case of the piezoelectric effect is shown in Figure 11, where the arrows represent dipole moments.[13]

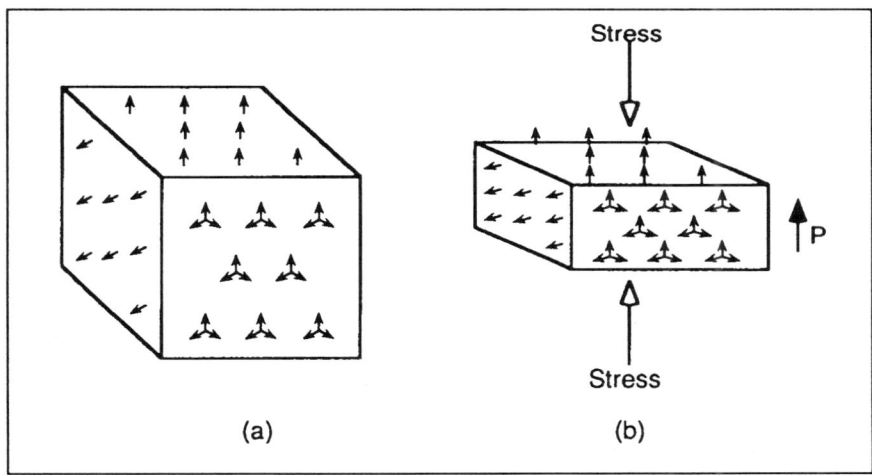

Figure 11. Arrows representing dipole moments illustrate how stress and polarization interact in piezoelectric materials: (a) unstressed isotropic material, (b) polarization P, develops as a result of stress applied to the material. (After Kittel reference 30).

SAW DEVICE CONSIDERATIONS

Figure 12a illustrates how piezoelectric materials are used to produce surface acoustic waves. Interdigital transducers (IDT's) are vapor deposited onto a substrate, with the period of the fingers equaling one acoustic wavelength. An RF signal is then sent to the IDT's where the electromagnetic signal interacts with the piezoelastic substrate. The interaction causes the substrate to become distorted by alternately straining then relaxing (Figure 12b), which propagates a surface acoustic wave across the substrate.[33]

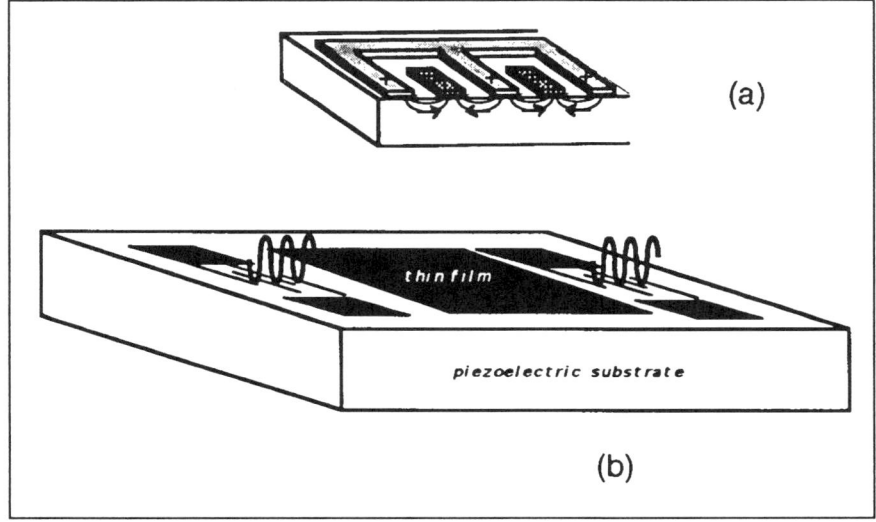

Figure 12. Surface acoustic wave generation on a piezoelectric substrate: (a) cut-away view of interdigital fingers showing electromagnetic interaction with piezoelectric substrate, which alternately distorts and relaxes the substrate, (b) mass sensing device showing surface acoustic wave propagation. (After reference 23).

The properties of the surface acoustic wave are critically dependent on crystal structure, cut and propagation direction, temperature and strain [34, 35]. In particular, SAW's are extremely sensitive to surface properties and conditions. They therefore make an excellent sensor for surface mass loading, yet properties which affect the sensor (such as temperature) must be carefully controlled.

SAW MASS SENSING

For sorption experiments, the SAW device is setup as a feedback element of an oscillator (Figure 13). The SAW device acts as sensitive microbalance, with sensitivity in the femtogram/cm^2 range.[10] (In more familiar units, this is equivalent to a surface density on the order of mg/km^2). In a film coated device, surface acoustic waves are not only sensitive to mass loading, but to film stiffness and conductivity changes as well.[9] This allows them to be utilized as sensitive surface reaction sensors as well. This novel characterization method may be an exciting new tool in the membrane field.

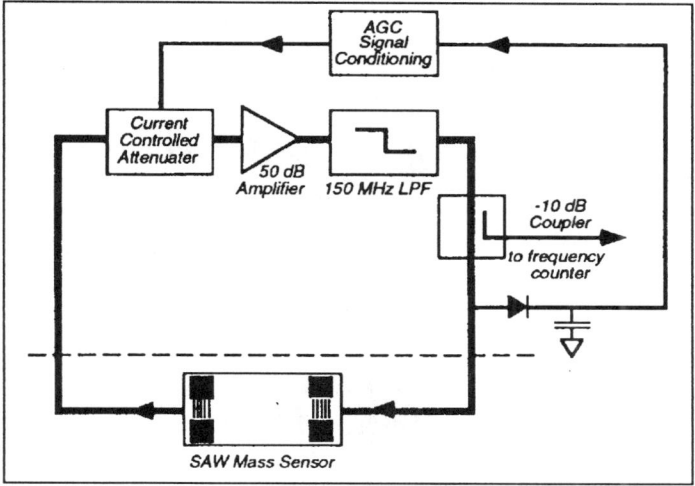

Figure 13.   SAW mass sensing electronics.

TRANSPORT MEASUREMENTS

Single-gas permeabilities of uncoated supports and supported membranes were measured as a function of pressure for He and N_2. The membrane, supported on a porous alumina tube, was mounted in a gas flow cell so compression at the tube ends into viton gaskets prevented gas bypassing. A mass flow controller was used to regulate the flow of the room temperature gases into the annulus of the membrane. Then the pressure drop (at steady-state mass flow rate) across the membrane was measured.

EXPERIMENTAL

In the sol-gel technique, metal alkoxides are reacted together to produce a gel which may be coated on substrates, or dried into a film directly. The advantages of this technique are (1) substitution of different metals in the alkoxides to produce various reactive sites or properties (2) control of the structure through control of the ligands, pH, water ratios, solvents, aging (3) a variety of deposition techniques including spin or dip coating and CVD.

Figure 14 shows schematically how the aluminosilicate sols were prepared: mix tetraethylorthosilicate (TEOS) and ethanol (molar ratios of 1TEOS:4ETOH) with an acid catalyst (HCl) in a round bottom flask; add aluminum tri(sec)butoxide (ASB) and reflux overnight at 353 K; then add water to form the gel. The bulk material was dried rapidly (at 413 K) after gellation, while the corresponding film samples were spun onto a SAW device at $t/t_{gel} = 0.5$ and 2000 rpm.

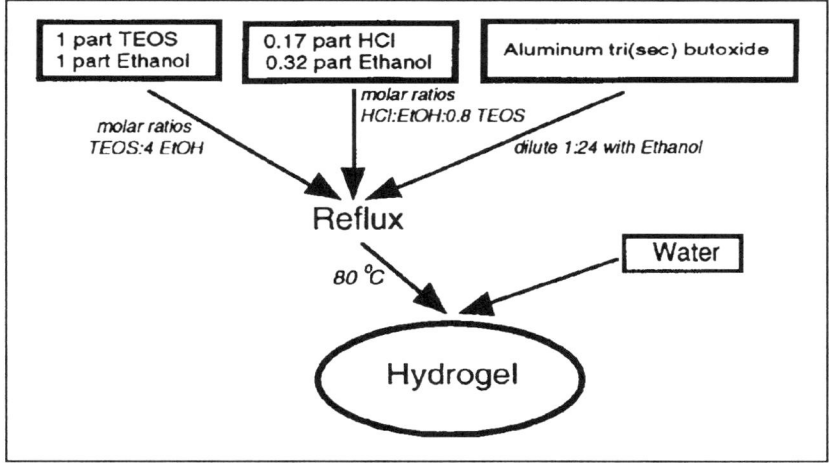

Figure 14.   Sol-gel synthesis scheme of aluminosilicate.

The SAW substrates are made from a piezoelectric ST-cut quartz wafer with Cr-Au evaporated interdigital transducer fingers at a 32 mm period. The device measures 8x11 mm with an active area $\approx$20 mm^2.

Dense silica particles with a diameter of $\approx$200 Å were prepared via a Stöber method by reacting TEOS in an ethanol solution with saturated ammonium hydroxide.[36] When formed into a film, these particles form mesoporosity with pore dimensions of 20 to 100 Å. Films were dried at 353 K for 15 minutes, then fired at 673 K for 15 minutes before analysis. After mounting in the sample cell, the films were outgassed at 443 K for 2 hours under vacuum. Adsorption measurements were made using nitrogen at 77 K. Standard gas adsorption data reduction techniques (BET, Kelvin and Dubinin-Radushkevich) were used (when applicable) to calculate surface area, pore volume, and pore size distribution in the films.

Permeability experiments were performed on films dip-coated onto asymmetric alumina tubes (macroporous outside and 40 Å pores inside). Prior to, and after coating, the alumina tubes were calcined at a rate of 1°C/ min up to 400 °C, held for ten minutes, then cooled to room temperature. The tubes were mounted onto a linear translation stage in a dry box, lowered and raised into/out of the sol at a rate of 20 cm/min, with a hold time of 100 seconds, and a drying period of 15 minutes. The substrates, and companion <100> single crystal silicon wafers for ellipsometry analysis, were coated at t/t_{gel} = 0.5.

RESULTS

The efficacy of the SAW technique in the experimental setup was observed by characterizing an uncoated sample, and a sol-gel prepared silicate film which has been studied extensively. A nitrogen sorption isotherm (adsorption and desorption branches) of an uncoated SAW device is shown in Figure 15 as well as a "BET plot" of the low relative pressure portion of the data. The Type 2 isotherm; relatively low uptake and lack of hysteresis, is typical of a non-porous material. The high precision of the uptake measurement is illustrated by the excellent fit of the BET equation despite the fact that the active area of the SAW device is only 20 mm^2, which implies that the volumetric uptake corresponding to a monolayer is only ~10^{-5} cm^3. The calculated surface area of 1.4 m^2/m^2 is only slightly higher than theoretical (1 m^2/m^2), a result of surface roughness. The measured BET C constant of 46 is in good agreement with previous nitrogen adsorption results on non-microporous silica (the SAW device has an oxide surface layer).[37] The ability of the SAW/ASAP to accurately control pressure and measure uptakes at relative pressure near unity is illustrated by the rapid increase in the isotherm and the lack of hysteresis.

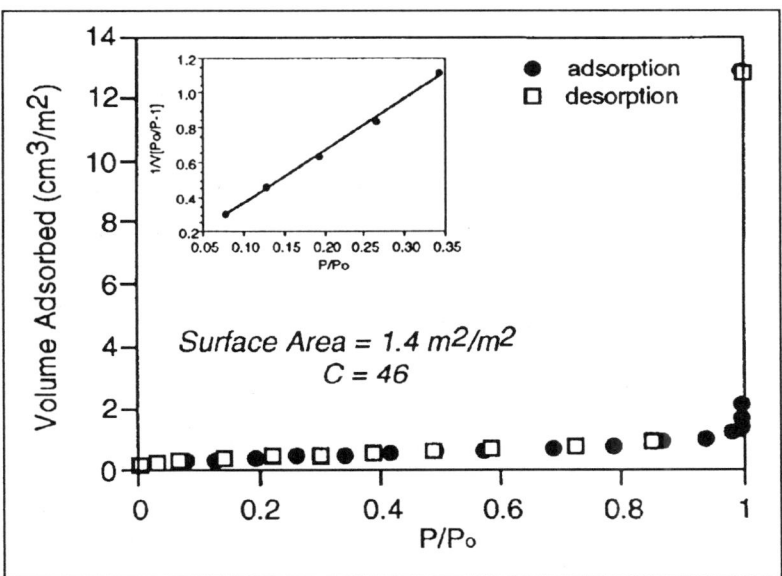

Figure 15.   Adsorption/desorption isotherms and BET plot for uncoated SAW device.

In contrast to the uncoated SAW device, results for the mesoporous Stöber film are shown in Figure 16. This isotherm is a Type 4 isotherm with considerable hysteresis. BET analysis yields a surface area of 20.2 m^2/m^2 and a BET C constant of 52. The small inset graph in Figure 16 shows the pore size distribution calculated from the desorption branch of the isotherm using the Kelvin equation. The hydraulic radius (twice the pore volume to surface area) is 59 Å which is in good agreement with the particle radius of ~10 Å and the maximum in the pore size distribution ($\approx$50 Å).

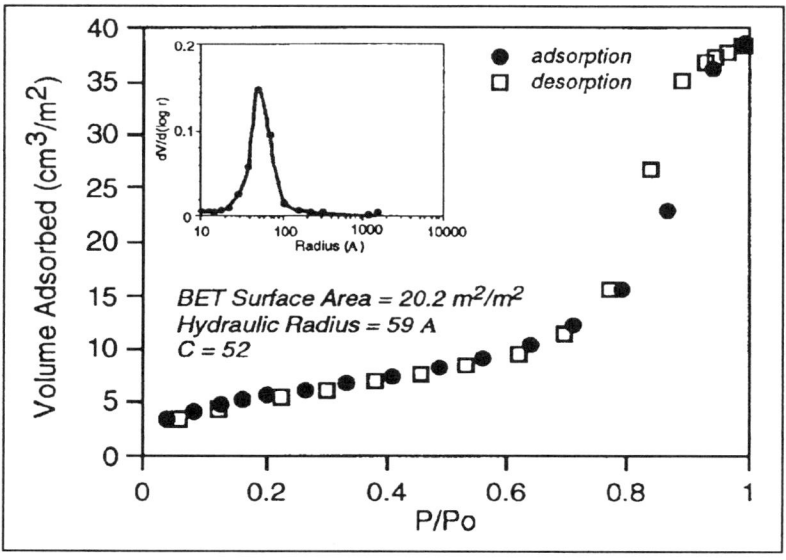

Figure 16. Adsorption/desorption isotherms and pore size distribution plot for Stöber silica coated SAW device.

Investigation of the effect of various water ratios on the 1:1 aluminosilicate composition on pore properties of the membranes were performed to determine whether they mimic that of the bulk prepared materials. Bulk and films of the 1:1 aluminosilicate composition with varying water to silicon ratios were prepared and characterized using SAW and permeability measurements. The surface areas of the bulk material were 428 m^2/g, 168 m^2/g and $\approx$1 m^2/g for the water to silicon ratios of 40, 60 and 100 respectively. The surface area of the films were measured and all found to be around 2 m^2/m^2 (2.16 m^2/m^2, 2.15 m^2/m^2 and 1.9 m^2/m^2 for the water to silicon ratios of 40, 60 and 100 respectively).

Figures 17, 18 and 19 are plots of the isotherms for the bulk/film 1:1 aluminosilicate composition with water to silicon ratios of 40, 60, and 100 respectively. The largest uptake in the bulk and film (water to silicon of 40) corresponds to the highest surface area/pore volume material, down to the basically nonporous material (water to silicon ratio 100).

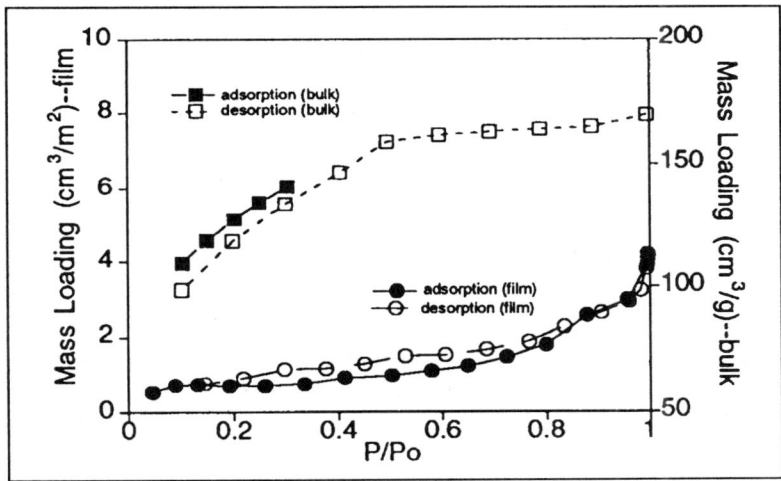

Figure 17. Bulk and film isotherms for 1:1 aluminosilicate with water to silicon ratio of 40.

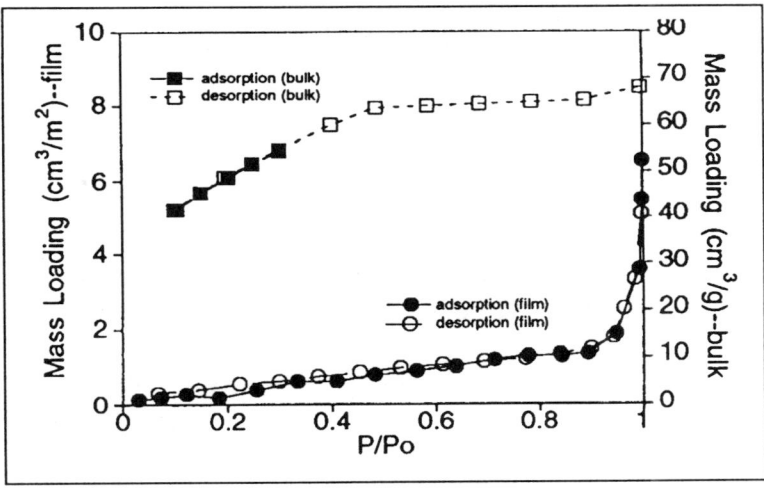

Figure 18. Bulk and film isotherms for 1:1 aluminosilicate with water to silicon ratio of 60.

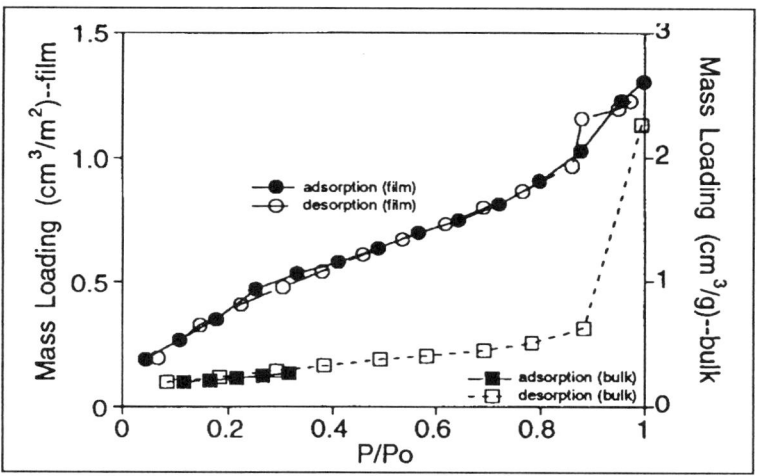

Figure 19.   Bulk and film isotherms for 1:1 aluminosilicate with water to silicon ratio of 100.

The isotherms for the bulk powders are similar to Type 1 isotherms, indicative of physical adsorption of microporous powders whose pore size does not exceed a few adsorbate molecular diameters. The shape of the isotherm is distinct since the gas molecule inside the small diameter pore is affected by the pore wall potential, which enhances the quantity adsorbed at low relative pressures. At higher relative pressures the pores are filled by adsorbed or condensed adsorbate leading to a plateau, with little or no external surface for additional adsorption.[39]

The film isotherms tend to be a hybrid, that is, not entirely Type 2 (nonporous). Type 2 isotherms frequently occur in adsorption on nonporous surfaces, with the inflection point usually occurring near the completion of the first adsorbed monolayer. At higher relative pressures, more layers are completed, until at saturation there becomes an infinite number of adsorbed layers, exemplified as a distinct rapid rise in the isotherm. As in the bulk material, the lower water ratios tend to be less Type 2, with the water to silicon ratio of 100 having very little adsorption. The membrane differences are further studied by permeability experiments of the films on alumina substrates.

The permeability results are presented as permeance, which relates a measured permeation flux to an overall driving force (e.g. the inverse of the total transport resistance).

$$J = F_T \Delta P \tag{4}$$

where J is the flux, F_T is the permeance of support and coatings, and ΔP is the pressure across support and coatings. The permeance results for these coatings are generally independent of pressure, consistent with Knudsen transport. In the Knudsen regime the separation factor is related to the inverse square root of the molecular weights of the gases, which for He/N$_2$ is 2.65. Figures 20, 21 and 22 show the permeance of the 1:1 aluminosilicate with water to silicon ratios of 40, 60 and 100 respectively. For the water to silicon ratio of 40 (Figure 20), two aluminosilicate coatings resulted in a nominal change in the He/N$_2$ separation factor of 2.47 for the uncoated tube to 2.23, with a corresponding reduction in permeance of about a factor of 10.

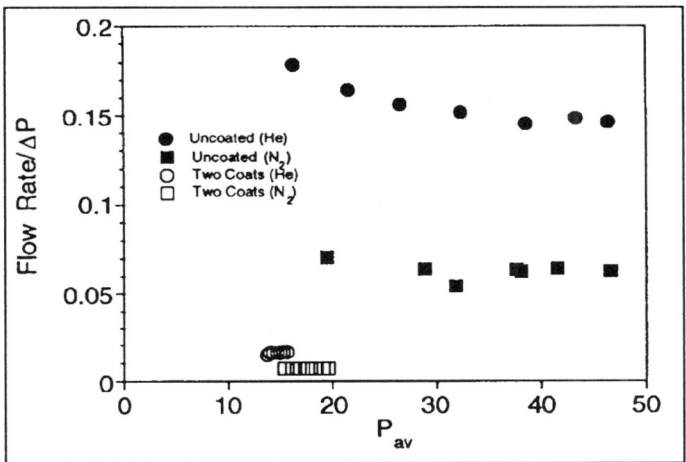

Figure 20.   Permeance of 1:1 aluminosilicate (water to silicon ratio of 40) coated on alumina tube.

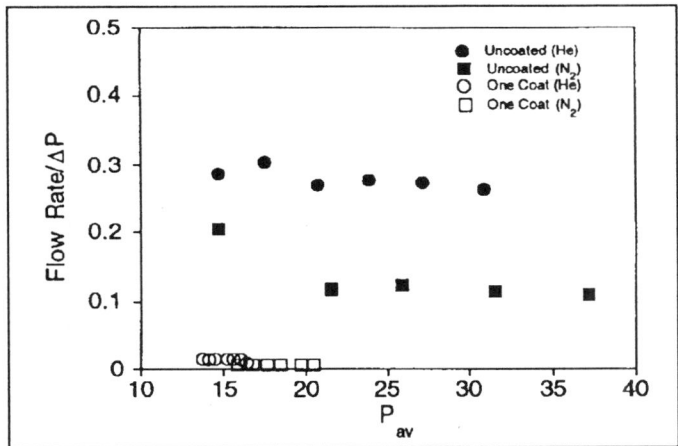

Figure 21.   Permeance of 1:1 aluminosilicate (water to silicon ratio of 60) coated on alumina tube.

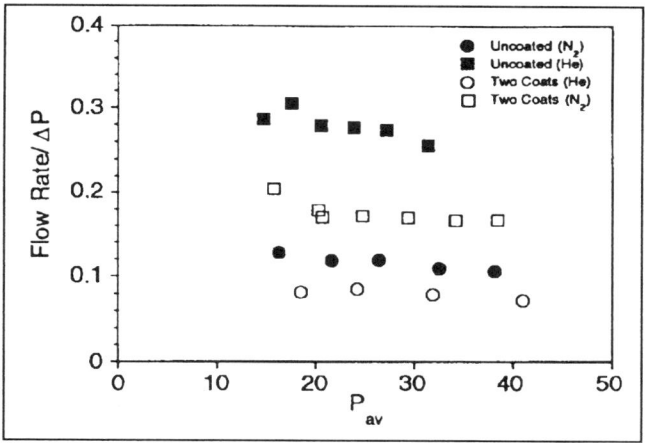

Figure 22.   Permeance of 1:1 aluminosilicate (water to silicon ratio of 100) coated on alumina tube.

One coating of the aluminosilicate with water to silicon ratio of 60 (Figure 21), resulted in a (again nominal) change in the He/N_2 separation factor of 2.13 to 2.42, and a reduction in the permeance by a factor of nearly 20. From the bulk as well as the SAW data, this type of trend was expected, for increasing porosity (pore volume) with decreasing water to silicon ratios. Figure 22 shows the 1:1 aluminosilicate with a water to silicon ratio of 100. For this sample, a small change in separation factor from 2.40 in the uncoated tube to 1.99 for the tube with two coats. The permeance reduction of <5, as well as the reduction in the selectivity, suggests that the coating might have been defective. The reduction in permeance indicates that the aluminosilicate material is coating the porous substrate, yet the nominal changes in the selectivity factors indicate that the pore size has not yet been dramatically reduced or that surface diffusion has become more prevalent. This data encourages further investigation to quantitatively determine differences in the coatings using gases of different sizes, and experimenting at different temperatures to determine the surface diffusion contribution.

CONCLUSIONS

The sol-gel technique has been demonstrated as a versatile method for preparing membranes, and in the aluminosilicate system, one in which the membrane properties may be effectively modified. The aluminosilicate membranes displayed slight differences with changes in processing conditions, with trends not dissimilar (but on a much smaller scale) to those found in the bulk materials. The aluminosilicate membranes, although not as distinctly microporous as their bulk counterparts when probed with nitrogen (by SAW), may more clearly display microporosity when probed with another gas, as has been shown in silicate membranes. Continued research will concentrate on identifying specific factors which control pore size formation and porosity in these aluminosilicate membranes.

The characterization methods used for studying the membranes have shown great research potential. Incorporation of the SAW technique into a commercial volumetric gas sorption instrument allows effective characterization of membranes, and accesses the enormous data analysis resources commonly used in bulk material characterizations. This inexpensive addition to the ASAP 2000 should also promote the industry-wide use of surface acoustic waves as an economical membrane characterization tool.

REFERENCES

[1]   J. Haggin, "Membrane Technology has Achieved Success, Yet Lags Potential," *C&EN*, October 1, 22-26 (1990).

[2]   K.K. Chan and A.M. Brownstein, "Ceramic Membranes-Growth Prospects and Opportunities," *Ceramic Bulletin* 4(70), 703-707 (1991).

[3]   J.B. Wachtman and L.M. Sheppard, "Inorganic Thin Films: Expanding Applications," *Ceramic Bulletin* 1(68), 91-95 (1989).

[4]   J.J Fripiat, A. Leonard and J.B. Uytterhoeven, "Structure and Properties of Amorphous Silicoaluminas. II. Lewis and Brönsted Acid Sites," *J. Phys. Chem.* 69(10), 3274-3279 (1965).

[5]   J.C. Pouxviel and J.P. Boilot, "Growth Process of Al_2O_3-SiO_2 Gels," Presented at the Conference on Ultrastructure Processing of Ceramic, Glasses and Composites, San Diego, CA February (1987).

[6]   D.W. Hoffman, R. Roy and S. Komarneni, "Diphasic Xerogels, A New Class of Materials: Phases in the System Al_2O_3-SiO_2," *J. Am. Ceram. Soc.* 67(7), 468-471 (1984).

[7]   S.L. Hietala, J.L. Golden, D.M. Smith and C.J. Brinker, "Anomalously Low Surface Area and Density in the Silica/Alumina Gel System," *J. Am. Ceram. Soc.* 72(12), 2354-2358 (1989).

[8]   A.S. Michaels and T.M. Meltzer (moderators) of Capitola Consultants, *Membrane and Separation Technology News*, April, 3 (1992).

[9]   J. Krim and V. Panella, *Characterization of Porous Solids II*, Rodriguez-Reinoso, Rouquerol, Sing, Unger, Eds.; p. 217, Elsevier Press: New York, (1991).

[10] G.C. Frye, A.J. Ricco, S.J. Martin and C.J. Brinker, "Characterization of the Surface Area and Porosity of Sol-Gel Films Using SAW Devices," *Materials Research Society Proceedings*, C.J Brinker, D.E. Clark, D.R. Ulrich, Eds., pp. 349-354, Materials Research Society, Pittsburgh, PA (1988).

[11]   W.G. Fahrenholtz, S.L. Hietala, P.P. Newcomer, N.R. Dando, D.M. Smith and C.J. Brinker, "Effect of Physical Structure on the Phase Development of Aluminosilicate Gels," *J. Am. Cer. Soc.* **74**(10), 2393-2397 (1991).

[12]   S.L. Hietala, D.M. Smith, C.J. Brinker, A.J. Hurd, A.H. Carim and N.R. Dando, "Structural Studies of Anomalous Behavior in the Silica-Alumina Gel System," *J. Am. Ceram. Soc.* **73**(10), 2815-2821 (1990).

[13]   C.J. Brinker and G.W. Scherer, S*ol-Gel Science: The Physics and Chemistry of Sol-Gel Processing,* Academic Press, New York, (1990).

[14]   S.S. Kistler, "Coherent Expanded Aeogels," *J. Phys. Chem.* **36**, 52-64 (1932).

[15]   S.J. Teichner, G.A. Nicolaon, M.A. Vicarini and G.E.E. Gardes, "Inorganic Oxide Aerogels," *Adv. Colloid Interface Sci.,* **5**, 245-273 (1976).

[16]   J. Fricke and G. Reichenauer, "Structural Investigation of SiO_2-Aerogels," *J. Non-Cryst. Solids,* **95 & 96**,1135-1142 (1987).

[17]   S.L Hietala, *Synthesis and Characterization of a Sol-Gel Prepared Aluminosilicate with Novel Properties,* M.S. Thesis, University of New Mexico (1990).

[18]   Lord Rayleigh, "On Waves Propagated Along the Plane Surface of an Elastic Solid," *Proc. London Math. Soc.* **17**, 4-11 (1885).

[19]   A.J. Devries and R. Adler, "Case History of a Surface-Wave TV IF Filter for Color Television Receivers," *Proc. of the IEEE* **64**(5), 671-676 (1976).

[20]   K. Shibayama, K. Yamanouchi, H. Sato and T. Meguro, "Optimum Cut for Rotated Y-Cut $LiNbO_3$ Crystal Used as the Substrate of Acoustic-Surface-Wave Filters," *Proc. of the IEEE* **64**(5), 595-597 (1976).

[21]   L.A. Coldren and H.J. Shaw, "Surface-Wave Long Delay Lines," *Proc. of the IEEE* **64**(5), 598-609 (1976).

[22]   I.M. Mason, E. Papadofrangakis and J. Chambers, "Acoustic-Surface-Wave Disk Delay Lines," *Proc. of the IEEE* **64**(5), 610-612 (1976).

[23]   H. Wohltjen and R. Dessy, "Surface Acoustic Wave Probe for Chemical Analysis I. Introduction and Instrument Description," *Analytical Chem.* **51**(9) 1458-1464 (1979).

[24]   S.J. Martin, G.C. Frye, A.J. Ricco and T.E. Zipperian, "Measuring Thin film Properties Using SAW Devices: Diffusivity and Surface Area," *IEEE Ultrasonics Symposium,* 563-567 (1987).

[25]   T.M. Reeder and D.E. Cullen, "Surface -Acoustic-Wave Pressure and Temperature Sensors," *Proc. IEEE* **64**(5), 754-756 (1976).

[26]   S.J. Martin, A.J. Ricco, D.S. Ginley and T.E. Zipperian, "Isothermal Measurements and Thermal Desorption of Organic Vapors Using SAW Devices," *IEEE Transaction on Ultrasonics, Ferroelectrics and Frequency Control,* **UFFC-34**(2), 142-147 (1987).

[27]   A.J. Ricco and S.J. Martin, "Thin Metal Film Characterization and Chemical Sensors: Monitoring Electronic Conductivity, Mass Loading

and Mechanical Properties with Surface Acoustic Wave Devices," *Thin Solid Films* **206**, 94-101 (1991).

[28] L. Sun, R.C. Thomas, R.M. Crooks and A.J. Ricco, "Real-Time Analysis of Chemical Reactions Occurring at a Surface-Confined Organic Monolayer," *J. Am. Chem. Soc.* **113**, 8550-8552 (1991).

[29] B.A. Auld, *Acoustic Fields and Waves in Solids, vol. I*, Robert E. Krieger Publishing Company, FL (1990).

[30] C. Kittel, *Introduction to Solid State Physics, 5th ed.*, John Wiley & Sons, NY (1976).

[31] D. Penunuri, K.H. Yen and R.B. Stokes, "Microwave Surface Acoustic Wave Devices," Chap 6 in *Handbook of Microwave and Optical Components, Volume 1: Microwave Passive and Antenna Components*, K. Chang, Ed., John Wiley & Sons, NY, (1989).

[32] B.A. Auld, *Acoustic Fields and Waves in Solids vol. II*, John Wiley & Sons, NY (1973).

[33] A.J. Slobodnik, Jr., "Surface Acoustic Waves and SAW Materials," *Proc. IEEE* **64**(5), (1976).

[34] A.R. West, *Solid State Chemistry and its Applications*, John Wiley & Sons, NY (1984).

[35] M.B. Boisen and G.V. Gibbs, "Mathematical Crystallography: An Introduction to the Mathematical Foundations of Crystallography," *Reviews in Mineralogy, vol. 15*, P.H. Ribbe, Ed., Mineralogical Society of America, Washington, D.C. (1990).

[36] W. Stöber, A. Fink and E. Bohn, "Controlled Growth of Monodisperse Silica Spheres in the Micron Size Range," *J. Colloid and Interface Science* **26**, 62-69 (1968).

[37] S.J. Gregg and K.S.W. Sing, *Adsorption, Surface Area and Porosity, 2nd ed.*; Academic Press, San Diego, CA (1982).

[38] S. Lowell and J.E. Shields, *Powder Surface Area and Porosity; 2nd edition*, Chapman and Hall Ltd., New York, (1984).

[39] S.L. Hietala, D.M. Smith, V.M. Hietala, G.C. Frye and S.J. Martin, "Pore Structure Characterization of Thin Films Using a Surface Acoustic Wave/Volumetric Adsorption Technique", accepted by *Langmuir.*

Computation of a Practical Index to Avoid Voltage Collapse in Electric Power Systems

RENÉ JEAN-JUMEAU

School of Electrical Engineering
Cornell University,
Ithaca, NY 14853

ABSTRACT

This report documents the development of a tool for predicting voltagecollapse in electric power systems. Voltage collapse is generally caused by either of two types of system disturbances: load variations and contingencies. A number of performance indices intended to measure the severity of the voltage collapse problem have been proposed in the literature. These performance indices however can not answer questions such as "Can the system withstand another 100 MVar increase on bus 11?" This report establishes a new performance index based on the center manifold voltage collapse model that provides a direct relationship between its value and the amount of load demands that the system can withstand before collapse. One of the features that distinguishes the proposed performance index from existing ones is its ability to answer questions such as "Can the system withstand a simultaneous increase of 70 MW on bus 2 and 50 Mvars on bus 6?". This feature makes the performance index readily interpretable by operators, making it more practical than existing ones to assess a potential voltage collapse. A theoretical basis stemming from bifurcation theory for the proposed technique is given.

INTRODUCTION

Several recent power system blackouts were related to voltage collapse. This phenomenon is characterized by a slow variation in the system operating point in such a way that voltage magnitudes at load buses gradually decrease until a sharp, accelerated change occurs. Voltage collapse has been especially experienced by heavily loaded power systems subject to an increase in load demands. There has been a wide concensus that as power systems operate under increasingly stressed conditions, the ability to maintain voltage stability to avoid collapse becomes a serious concern.

Among several examples of voltage collapse, the recent voltage collapse in Japan in 1987 [1] was due to load variations and the voltage collapse in Sweden in 1982 [2] was caused by a contingency. Voltage collapse due to contingencies has been studied by several researchers, see for example [3–9], where the key issues are the feasibility of the stable equilibrium point after contingency and the estimate of its stability region (region of attraction). Voltage collapse has also been attributed to a lack of reactive power support, which can be equivalently regarded as due to increases in load demand [10].

A number of performance indices intended to measure the severity of the voltage collapse problem have been proposed in the literature. Among them, the minimum singular value in [11], later pursued in [12], and the condition number in [13] of system Jacobian intend to provide some measure of how far the system is away from the point at which the system Jacobian becomes singular. The performance index proposed in [14] and [15] is based on the the angular distance between the current stable equilibrium point and the closest unstable equilibrium point in a Euclidean sense. The performance index proposed in [16,17] measures the energy distance between the current stable equilibrium point and the closest unstable equilibrium point using an energy function. These performance indices can be viewed as providing some meaure relative to the "distance" between the current operating point and the bifurcation point. However, all these performance indices are defined in the state space of power system models. Thus, these performance indices cannot answer questions such as "Can the system withstand a 100 MVar increase on bus 11?" or "Can the system withstand a simultaneous increase of 70 MW on bus 2 and 50 Mvars on bus 6?".

After over two decades of study of occurences of actual voltage collapses and consideration of numerous approaches, it is now generally accepted that volatge collapse can be the result of either *dynamic* instabilities or a *static* loss of stability. In the first case the system is unable to control an unusual sequence of events—contingencies, tripping circuit breakers, operation of OLTCs, etc. This directs the system operating point along a trajectory which leads it out of its stablility region. In the second case, the system continually operates near a stable equilibrium or operating point, but loses stability as the region of attraction locally disappears, typically under heavily loaded conditions. This is the class of problems considered here. It has been noted that, in addition to the preceding scenarios, the dynamic events may also create conditions under which the static problem appears. Therefore, as Kwatny emphasizes in [18], "...static bifurcations are important enough... that it is appropriate to develop the conceptual and computational means to analyse systems of realistic scale." We thus report theoretical and practical results to this effect.

It should be stressed that one basic requirement for useful performance indices is their function to reflect the degree of *direct* mechanism leading the underlying system toward an undesired stage. In the context of voltage collapse in power systems, it is expected that any performance index has the ability to

1.   measure the amount of load increase that the system can tolerate before collapse (when the underlying mechanism of collapse is due to load variations), or
2.   assess whether the system can sustain a contingency without collapse (when the underlying mechanism of collapse is caused by a contingency) and measure the severity of the contingency.

The existing performance indices, however, generally do not exhibit any obvious relation between their value and the amount of the underlying mechanism that the system can tolerate before collapse. We assert that, in order to provide a direct relationship between its value and the amount of load increases that the system can withstand before collapse, the performance index must be developed in the parameter space (i.e. the space of load demands) rather than the state space where the existing performance indices were developed.

In this report we present a new performance index that provides a direct relationship between its value and the amount of load variations that the system can withstand before "static" collapse. The new performance index has the following distinguishing features:

•   it is more practical than existing ones in terms of its ability to provide a direct relationship between its value and the amount of parameter variation that the system can withstand before collapse. More specifically, it is readily interpreted to operators of power systems to answer questions such as "Can the system withstand a simultaneous increase of 70 MW, 40 Mvars on bus 2, 100 MW on bus 5 and 50 Mvars on bus 16?".
•   it provides useful information as to how to derive load-shedding schemes to avoid voltage collapse.

The new performance index as well as the voltage collapse margin presented in this paper have been tested on several power systems. Simulation results on the IEEE 39-bus and the Tai-power 234-bus system are presented with promising results.

PRELIMINARIES

In general, a power system can be described by

$$\dot{x} = f(x, \lambda) , \tag{1}$$

where $x \in R^n$ is a state vector including bus voltage magnitudes and angles, $\lambda \in R^m$ is an m-dimensional real parameter vector representing real and reactive power demands at each load bus. The parameter vector λ is subject to variation (due to load variations) causing a structure change in the load flow solutions of (2). For instance, the number of load flow solutions may change as λ varies. If the parameter λ varies slowly or quasi-statically with respect to the dynamics of (1), then the power system can be appropriately modeled by

$$0 = f(x, \lambda) . \tag{2}$$

We are interested in the situation when the system (2) is the so-called (one-parameter) Decoupled Parameter-dependent Nonlinear (DPDN) dynamical system. In power system applications, such a dynamical system is a system together with one of the following conditions:

1. the reactive (or real) power demand at one load bus varies while the others remain fixed,
2. both the real and reactive power demand at a load bus varies and their variations can be parameterized. Again the others remain fixed,
3. the real and/or reactive power demand at some collection of load buses varies and their variations can be parameterized while the others are fixed.

Algebraically, they can be written:

$$f(x, \lambda) = f(x) - b\lambda, \tag{3}$$

and their equilibria are the solutions of

$$f(x) - b\lambda = 0, \tag{4}$$

where $\lambda \in R$, $b \in R^n$ and $f: R^n \to R^n$ for an n-dimensional system.

Typically, the power system (2) operates at a stable load flow solution. In [19,20], Dobson et al. have linked the power system voltage collapse phenomenon to the disappearance of the stable EP in a saddle node bifurcation

(SNB) as the parameter λ passes through a critical value λ^*. In the terminology of nonlinear dynamical systems, λ^* is called the bifurcation value and (x^*, λ^*) is called the saddle node bifurcation point (SNBP) bifurcation point. Figure 1 recalls these concepts. The center manifold voltage collapse model proposed in [19] was based on the saddle-node bifurcation and its consequent dynamics. It has been shown in [20] that the model encompasses many existing models used to explain voltage collapse such as the multiple load flow model, the load flow feasibility model, the static bifurcation model, the singular Jacobian model and the system sensitivity model. Physical explanations of this center manifold voltage collapse model can be found in [21]. In the next section, we will present a new performance index based on the center manifold voltage collapse.

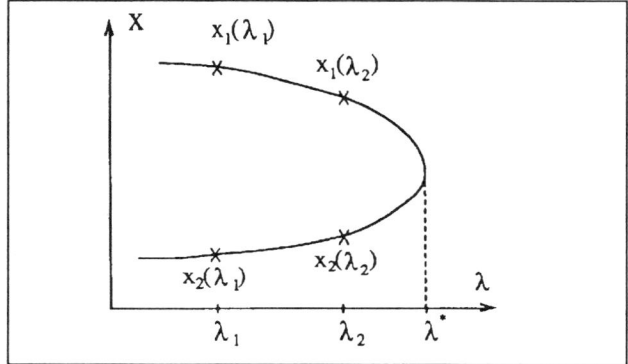

Figure 1.   Before a saddle-node bifurcation there is a pair of load flow solutions x_1–x_2. As parameter λ varies toward its bifurcation value λ^*, this pair of load flow solutions gets closer to each other and finally, x_1 and x_2 coalesce, when $\lambda = \lambda^*$, to form an equilibrium point x^* whose corresponding system Jacobian has one zero eigenvalue and the real parts of other eigenvalues are negative.

THE SEYDEL TEST-FUNCTION t_s AND THE APPROXIMATE
BIFURCATION VALUE

Given a system of nonlinear equations dependent on the parameter λ defined by equation (4), in order to compute the bifurcation value and the state variables at the bifurcation point, we need a practical means of approximating the bifurcation value for the parameter.

In order to do this, we again refer to equation (4) and utilize a "test function" defined by Seydel in [22]. Let $J(x, \lambda)$ denote the Jacobian matrix of $f(x, \lambda)$:

$$J(x, \lambda) = \frac{\partial f}{\partial x}(x, \lambda).$$

Consider the following test-function [22],

$$t_s(x, \lambda) \triangleq e_l^T J(x, \lambda) v. \tag{5}$$

This test function has the following desired characteristics: (a) it is a function of x and λ and (b) it has a value of *zero* at the bifurcation value λ^* of (4).

Seydel suggests in [22] that the above test-function (5) is expected to be, but is not restricted to, a parabolic function symmetrical about the λ-axis. To evaluate the suggestion, we have run computer simulations on several systems using a tool based on a continuation method to find the exact bifurcation value (Figure 2). These simulation results suggest that:

1. the test-function can be approximately quadratic when the parameter variations only affect one equation (as with a scalar parameter present in a single equation);
2. the test-function can be best fit by a quartic model when the variations affect several equations (as with a parameter vector or a scalar parameter present in several equations).

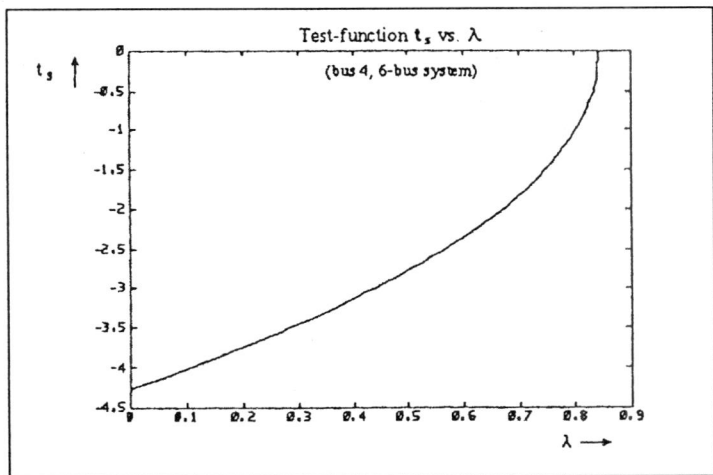

Figure 2. Test-function t_s obtained for a 6-bus network with respect to load increase on bus 3.

This leads, after some algebraic manipulations, to the following expressions for the value λ^*:

$$\lambda^* \approx \lambda_1 - \frac{1}{2}\frac{t_s(x_1, \lambda_1)}{t_s'(x_1, \lambda_1)} \tag{6}$$

for the quadratic model, and

$$\lambda^* \approx \lambda_1 - \frac{1}{4}\frac{t_s(x_1, \lambda_1)}{t_s'(x_1, \lambda_1)} \tag{7}$$

for the quartic model.

In practice, in order to avoid the large amount of computation needed to evaluate the derivative t_s' exactly, an approximation of t_s' can be found via the finite difference as:

$$t_s' \approx \bar{t}_s' = \frac{t_s(x(\lambda_1 + \delta\lambda), \lambda_1 + \delta\lambda) - t_s(x_1, \lambda_1)}{\delta\lambda}, \tag{8}$$

for small $\delta\lambda$, where $x_1 = x(\lambda_1)$. Perhaps using x_1 as initial guess, an EP $x(\lambda_1 + \delta\lambda)$ is computed. From this, in conjunction with equation (6) or (7), we get an *estimate* of the bifurcation value λ^*. Let $\bar{\lambda}$ denote this estimate:

$$\bar{\lambda} = \lambda_1 - \frac{1}{c}\frac{t_s(x_1, \lambda_1)}{t_s'(x_1, \lambda_1)}, \tag{9}$$

where c=2 or 4 for the quadratic or quartic models respectively. The margin to this approximation is the value of the voltage collapse index (VCI) τ:

$$\tau = \bar{\lambda} - \lambda_1.$$

As an example of its effectiveness, see Figure 3 for a sample comparison between τ and the exact margin $\Delta\lambda$ on bus 4 of a 6-bus test system.

The computation required in the above procedure basically involves calculations of two EP solutions (the current one and a new one) and some matrix computations. This makes the proposed approximation much easier and cheaper to compute than running continuation methods or utilizing some other technique proposed in the literature which requires calculations such as eigenvalues, eigenvectors, singular values, energy functions or conditional numbers. In addition, the proposed approximation provides a direct measure regarding the amount of parameter increase that the system can withstand before reaching the saddle-node bifurcation point. This information is valuable, as it is quickly available before the computation of the exact saddle node bifurcation value takes place. This is especially relevant in the case of very large systems.

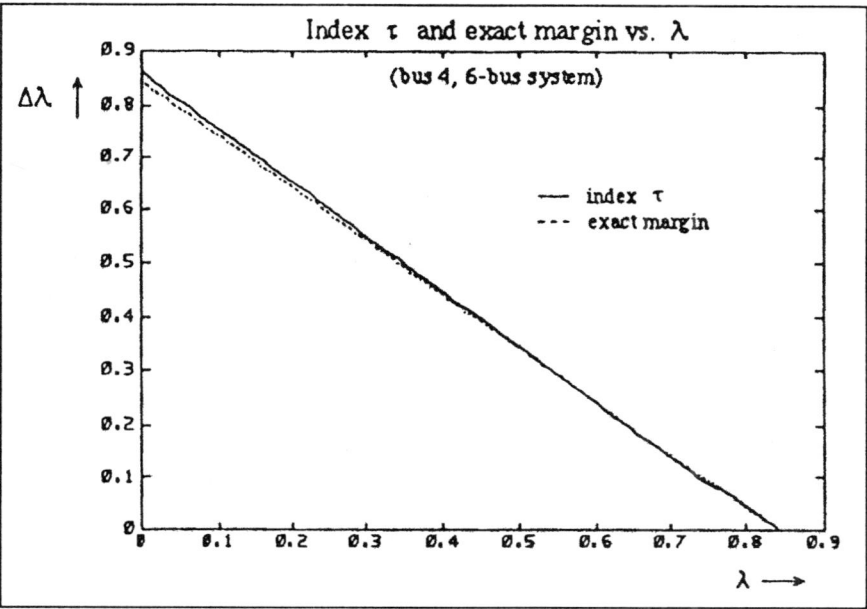

Figure 3. The estimated bifurcation value compared to the exact value λ^* in a test case.

ROUGH ESTIMATION OF STATE VARIABLE VALUES AT BIFURCATION

Due to the properties of SNBPs, near the bifurcation value λ^*, we can also assume a quadratic form for each component i of the state vector x and approximate the bifucation value in the state space with

$$\bar{x}_i = x_{1i} + 2(\lambda_1 - \bar{\lambda}) \cdot x'_{1i}. \tag{10}$$

See figure 4.

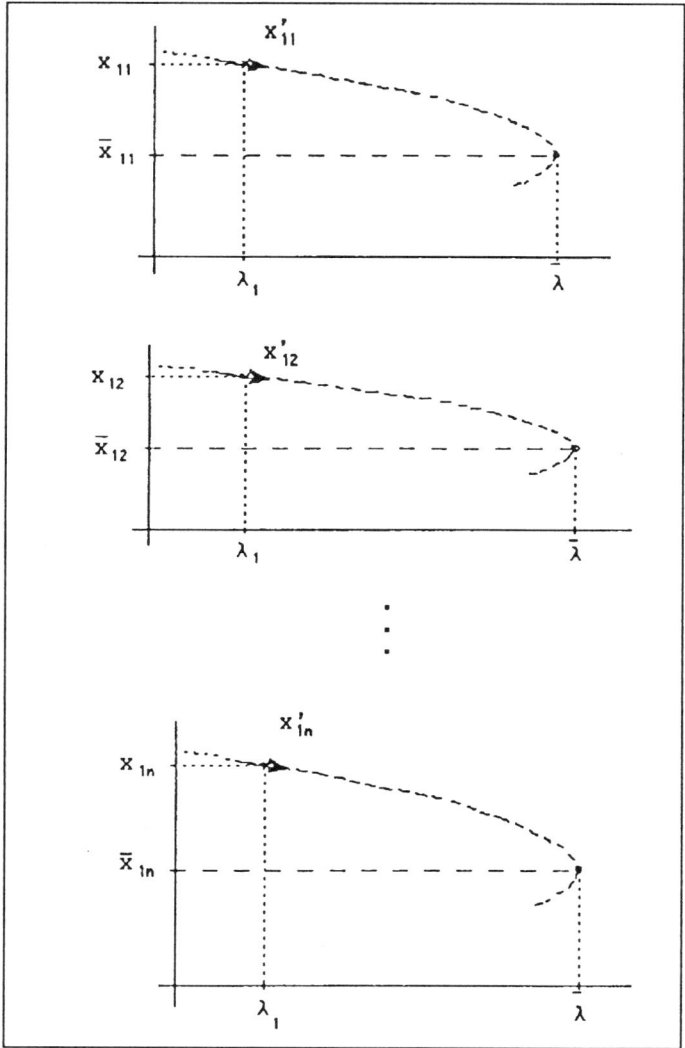

Figure 4.   Determination of the approximation $\bar{x}$ of the bifurcation point x^{*}.

Instead of computing the exact derivative vector x'_1, it is much more efficient to compute the secant approximation:

$$x'_1 \approx \frac{x(\lambda_1 + \delta\lambda) - x_1}{\delta\lambda}.$$

(11)

This can in turn be used to compute $\bar{x}$ with (10). Generally, far from the saddle-node bifurcation value of (4), the approximation $\bar{x}$ is not a good enough estimate of the bifurcation point x^*. However, it does provide us with an approximate state vector with which to accomplish a change of parameter that we prescribe in [23]. This yields a new parameter value

$$\bar{\kappa} = \frac{\bar{\lambda}}{b^T \bar{x}}. \tag{12}$$

which is to be used as an initial guess in solving for the approximate bifurcation point (x_2, λ_2). Note that we easily avoid convergence problems since, as we have shown in [23], the reformulated system is not ill-conditioned near (x^*, λ^*).

By solving the reformulated system as per our suggestion in [23], we get an equilibrium point solution x_2 close to x^*. Reverting back to the parameter λ, we get λ_2:

$$\lambda_2 = \bar{\kappa} b^T x_2, \tag{13}$$

yielding the SNBP approximation

$$(x_2, \lambda_2) \approx (x^*, \lambda^*).$$

NUMERICAL RESULTS

Here we take a look at different phases of the VCI determination process.

EFFECT OF THE CHANGE OF PARAMETERS

The bifurcation point is "moved" further along the solution curve of $F(x, \lambda) = 0$ when the parameter is changed from λ to κ via the relation

$$\lambda = \kappa b^T x = \kappa x_\iota,$$

where ι is the bus index.

This is supported graphically by figures 5 and 6.

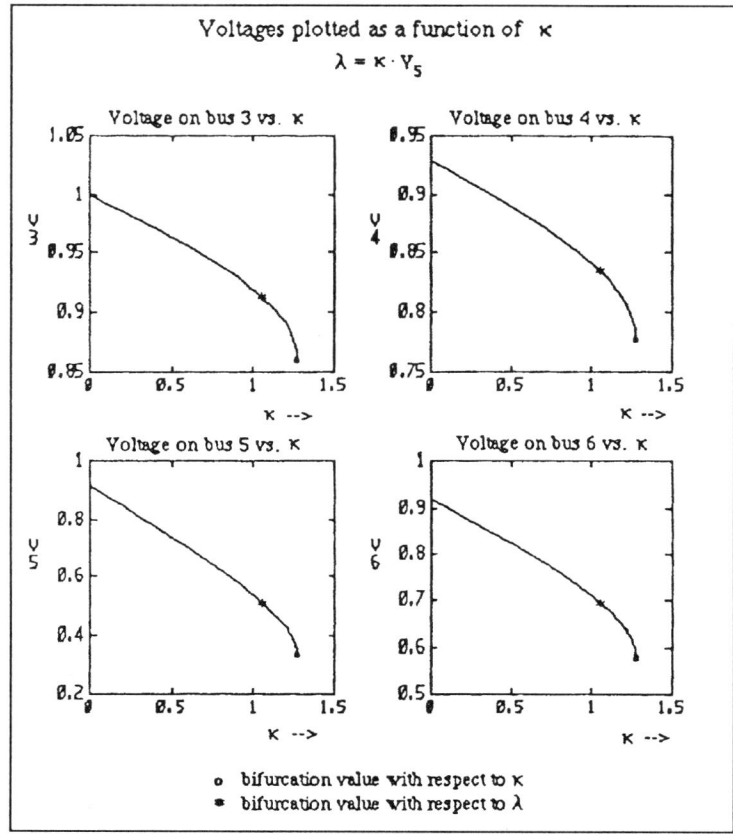

Figure 5.   Relative locations of the bifurcation points with respect to λ and κ when the parameter is changed.

We examine a practical 234-bus power system including 51 generator buses and 443 transmission lines. Suppose that we are interested in the system behaviour subject to reactive power variations on the 131st load bus labeled "5077". This analysis can be carried out by letting $b=e_{131}$ (the 131st standard unit vector) and finding the maximum value of λ satisfying the load flow equations: this correponds to a maximum or critical loading value with respect to reactive load demand on bus 5077. This critical λ is found to be $λ^* = 3.5789$ via an analytical procedure [22] and the voltage magnitude on the bus studied is $V_{5077} = x^*_{131} = 0.5520$. An attempt to run a load flow, from a "flat start" (i.e. with all voltage magnitudes at 1 p.u. and all angles at 0°, for $λ=λ^*$ using a modified Newton's method which largely avoids divergence [25] failed . Convergence is obtained easily, however, when the methodology proposed in section 4 is used with

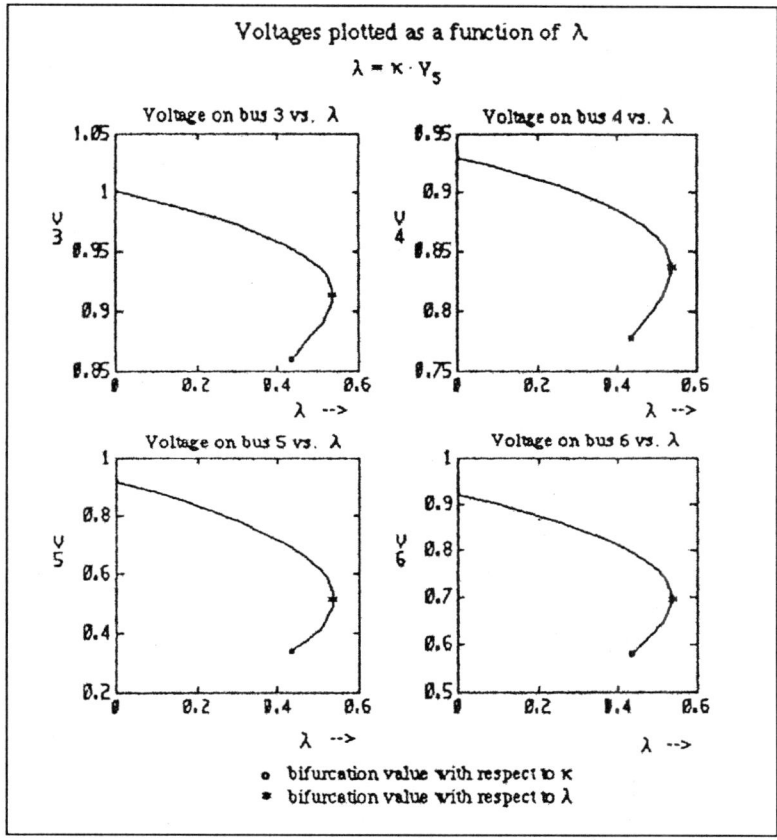

Figure 6. Relative locations of the bifurcation points with respect to λ and κ when the parameter is changed (Continued).

$$\kappa^* = \frac{\lambda^*}{b^T x^*} = 6.4840.$$

Tables 1 and 2 show how, although avoiding divergence, the modified Newton's method cannot converge in the first case, yet exibits normal quadratic convergence in the second. The total mismatch is defined as the sum of the absolute values of the real and reactive power mismatches. A tolerance of .000001 p.u. was used for the calculations. We note that, in this example, the degree of sparsity of the Jacobian matrix is unchanged, since "b transpose times b" only affects one diagonal value which is already non-zero. The computational complexity is here the same as that of a conventional load flow.

Table 1.   Load flow convergence at a critical point for a 234-bus network: λ as parameter.

parameter	λ	
value	$\lambda^* = 3.5789$	
	Iteration Number	Total Mismatch
	1	2039.73183
	2	534.75930
convergence	3	94.39755
process	4	12.98269
	5	10.6808
	6	3.61689
	7	2.07385
	8	2.53221
	Newton Error Problem with Convergence	

Table 2.   Load flow convergence at a critical point for a 234-bus network: κ as parameter.

parameter	κ	
value	$\kappa^* = 6.4840$	
	Iteration Number	Total Mismatch
	1	2043.39234
	2	527.50834
convergence	3	92.94211
process	4	11.12853
	5	0.59570
	6	0.00961
	7	0.00002
	8	0.00000
	Convergence attained	

THE TEST-FUNCTION t_s: MODEL VALIDATION

Using the methodology described in section 3, the test function is modeled as a polynomial function: second-order for variation on a single bus, fourth-order for multiple-bus parameter variation. This choice of a model is justified by the results obtained on various test networks and illustrated below (Figures 7, 8 and 9).

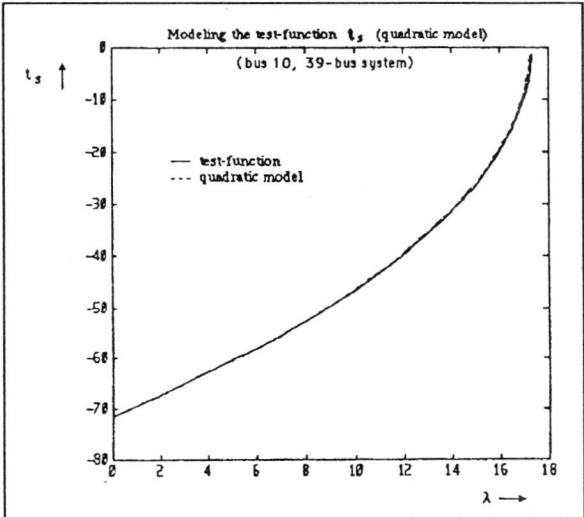

Figure 7.   Comparison of test-functions t_s with a quadratic model on a 39-bus test system.

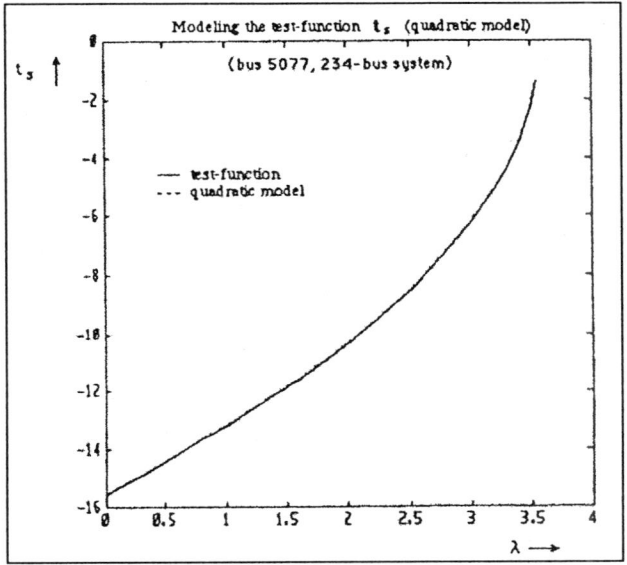

Figure 8.   Comparison of test-function t_s with a quadratic model on s 234-bus test system.

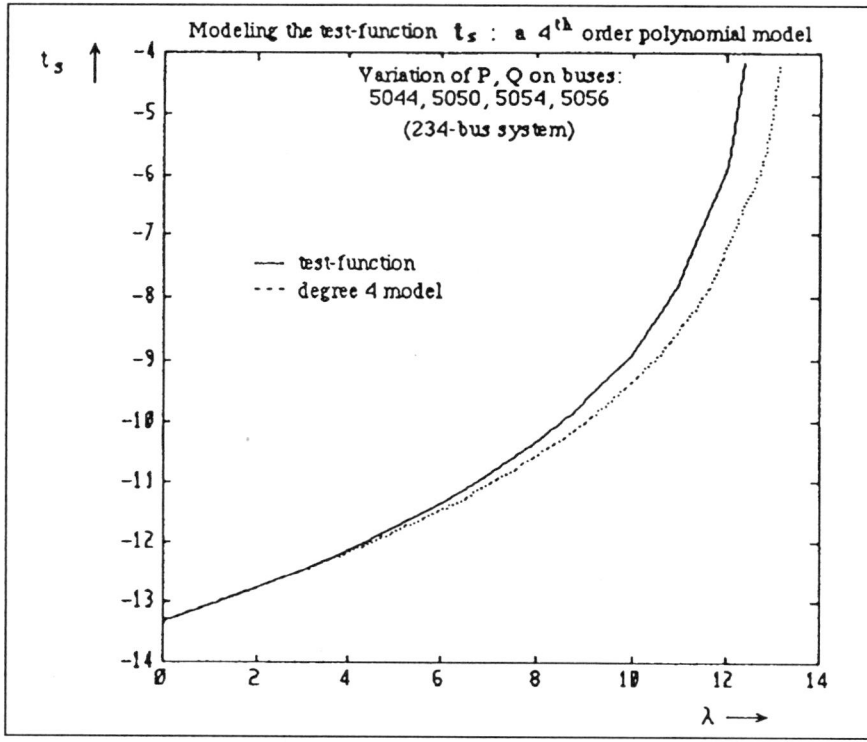

Figure 9. Relative aspects of the graphs of t_s and a fourth-order polynomial model.

A GENERAL CASE: MULTIPLE-BUS PARAMETERIZATION

Here, with the load flow equations in the form of equation (4), we may select a *cluster* of load buses for real and/or reactive load variation. Then,

$$b = \sum_{i \in J} \alpha_i e_i,$$

where $J \subset \{1,2,\ldots, n\}$, $\alpha_i \in \Re$ and the e_i are the ith standard unit vectors in $\Re^n$.

The test-function

In the multiple-bus case, a fourth degree polynomial better approximates the test-function. This is shown in the simultaneous plot of t_s and its quartic model in figure 9. The test-function t_s and the model are plotted when real and reactive loads are varied simultaneously on 4 load buses of the 234-bus system.

The estimate of the bifurcation point: $\bar{\lambda}$.

The resulting estimate $\bar{\lambda}$ of the SNBP is plotted in comparison to the exact value λ^* for both the 6-bus and the 234-bus systems in figure 10. Real and reactive power variations are considered on several load buses in each case.

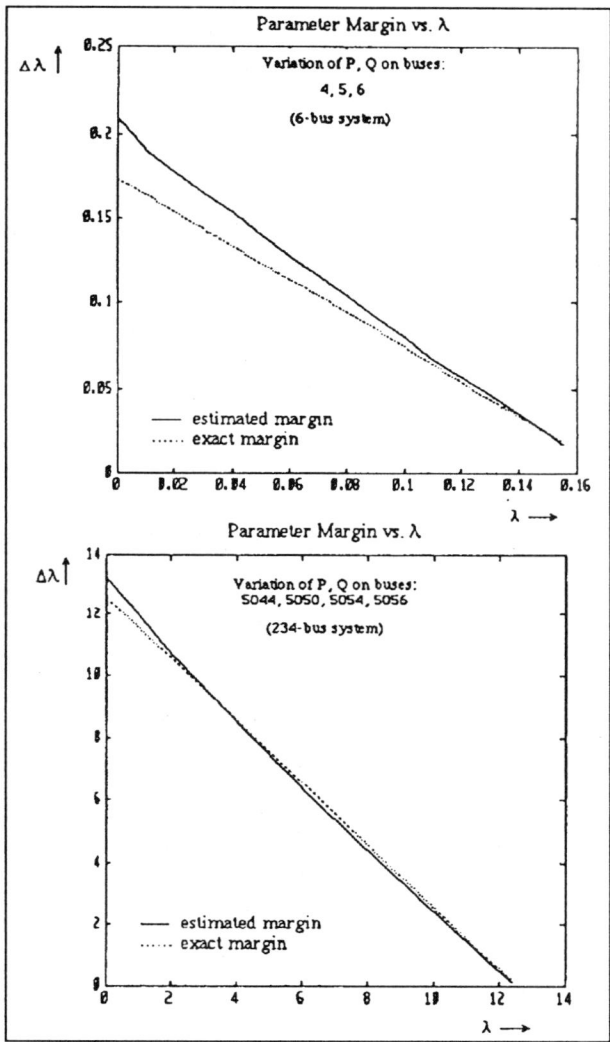

Figure 10.   Comparison of the estimated parameter margins τ with the exact margins $\Delta\lambda$ in the multiple-bus case.

CONCLUSIONS

We have presented a new performance index based on the center manifold voltage collapse model that provides a direct relationship between its
value and the amount of load demands that the system can further withstand
before collapse. This relation makes the performance index readily interpretable to operators, making it more practical than existing ones to assess
a potential voltage collapse. This new performance index will also allow us
to identify the "weak areas" subject to voltage collapse and to derive pre-
ventive controls for the problem. One of the features that distinguishes the
performance index developed in this paper from existing ones is that this
one is developed in the parameter space. As a contrast, the existing ones
were developed in the state space and can not answer questions such as
"Can the system withstand a 100 MVar increase on bus 11?" or "Can the system withstand a simultaneous increase of 70 MW on bus 2 and 50 Mvars on
bus 6?". We are encouraged by the promising simulation results on several
power systems. We hope to extend the performance index to more detailed
power system models.

As a great majority of performance indices for voltage collapse proposed
to date, the new performance index presented in this paper used the conventional power flow model to represent the system steady state. However,
this may not always be appropriate, in particular when the system is heavily
loaded. From a more practical viewpoint, there is a need to consider detailed steady state models for key system components critical to analysis of
voltage collapse such as generators, SVCs, under-load tap changers, and so
on. More research is needed as well in devising suitable load models for the
analysis of voltage collapse. Our ongoing research is oriented in those di-
rections.

REFERENCES

[1]   A. Kurita and T. Sakurai, "The power system failure on July 23, 1987
      in Tokyo," *IEEE Proceedings of the 27th Conference on Decision and
      Control*, Austin, TX, 2093–2097 (1988).

[2]   K. Walve, "Modelling of Power System Components at Severe
      Disturbances," *Proceedings of the 31st International Conference on
      Large High Voltage Electric Systems*, Paris, France, 2, paper 38-18
      (1986).

[3]   H.D. Chiang and F.F. Wu, "On Voltage Stability," *Proc. 1986 IEEE
      International Symposium on Circuits and Systems*, 3, 1339–1343
      (1986).

[4]   F. Mercede, J.C. Chow, H. Yan and R. Fischl, "A Framework to Predict
      Voltage Collapse in Power Systems," *IEEE Trans. on Power Systems*, 3
      (4), 1807–1813 (1988).

[5]   T. Van Cutsem, "Dynamic and static aspects of voltage collapse," *Proceedings: Bulk Power System Voltage Phenomena—Voltage Stability and Security*, EPRI Report EL-6183, 6-55–6-80 (1989).

[6]   R. A. Schleuter, A. G. Costi, J. E. Sekerke, H. L. Forgey, "Voltage Stability and Security Assessment," *EPRI Final Report* EL-5967 (1988).

[7]   C.C. Liu, K.T. Vu, "Analysis of Tap-changer Dynamics and Construction of Voltage Stability Regions," *IEEE Trans. on Circuits and Systems*, 36 (4), 575–590 (1989).

[8]   R. A. Schleuter, J.C. Lo, T. Lie, T.Y. Guo and I. Hu, "A fast accurate method for midterm transient stability simulation of voltage collapse in power systems," *IEEE Proceedings of 28th Conference on Control and Decision*, Tampa, FL., 340–344 (1989).

[9]   B.H. Lee, K.Y. Lee, "A Study on Voltage Collapse Mechanism in Electric Power Systems," *IEEE Trans. on Power Systems*, 6 (3), 966–974 (1991).

[10]   W.R. Lachs, "Insecure System Reactive Power Balance Analysis and Countermeasures," *IEEE Trans. on Power Apparatus and Systems*, 104 (9), 2413–2419 (1985).

[11]   A. Tiranuchit, R. J. Thomas, "A Posturing Strategy Against Voltage Instabilities in Electric Power Systems," *IEEE Trans. on Power Systems*, 3 (1), 87–93 (1988).

[12]   P.A. Löf, T. Smed, G. Andersson, D.J. Hill, "Fast Calculation   of   a Voltage Stability Index", *IEEE Trans. on Power Systems*, 7 (1), 54-64 (1992).

[13]   M.G. O'Grady and M.A. Pai, "Analysis of Voltage Collapse in Power Systems", *Proceedings of the 21st Annual North American Power Symposium*, Rolla, MO 151-160 (1989).

[14]   Y. Tamura, H. Mori, and S. Iwamoto, "Relationship Between Voltage Stability and Multiple Load Flow Solutions in Electric Power Systems," *IEEE Trans. on Power Apparatus and Systems*, 102 (5), 1115–1125 (1983).

[15]   Y. Tamura, "Voltage Instability Proximity Index Based on Multiple Load-flow solutions in Ill-conditioned Power Systems," *IEEE Proceedings of 27th Conference on Control and Decision*, Austin, TX, 2114–2119 (1988).

[16]   C.L. DeMarco, T.J. Overbye, "An Energy Based Security Measure for Assessing Vulnerability to Voltage Collapse", *IEEE Trans. on Power Systems*, 5 (2), 419-427 (1990).

[17]   T.J. Overbye and C.L. DeMarco , "Voltage Security Enhancement Using An Energy Based Sensitivity", *IEEE Trans. on Power Systems*, 6 (3), 1196–1202 (1991).

[18]   H.G. Kwatny, "Steady State Analysis of Voltage Instability Phenomena," *Proc. Bulk Power Systems Voltage Phenomena—Voltage Stability and Security*, EPRI Report EL-6183, 5-1–5-22 (1989).

[19] I. Dobson and H.D. Chiang, "Towards a Theory of Voltage Collapse in Electric Power Systems," *System and Control Letter*, 13 (3), 253–262 (1989).

[20] I. Dobson, H.D. Chiang, J.S. Thorp, L. Fekih-Ahmed, "A Model of Voltage Collapse in Electric Power Systems," *IEEE Proceedings of the 27th Conference on Control and Decision*, Austin, TX, 2104–2109 (1988).

[21] H.D. Chiang, I. Dobson, R.J. Thomas, J.S. Thorp, L. Fekih-Ahmed, "On the Voltage Collapse in Power Systems," *IEEE Trans. on Power Systems*, 5 (2), 601–611 (1990).

[22] R. Seydel, "Numerical Computation of branch points in Nonlinear Equations," *Numer. Math.*, 33, 339–352 (1979).

[23] R. Jean-Jumeau, H.D. Chiang, "Parameterizations of the Load-Flow Equations for Eliminating Ill-conditioning Load Flow Solutions", to appear in *IEEE Trans. on Power Systems* Fall (1992).

[24] H.D. Chiang, W. Ma, R.J. Thomas, and J.S. Thorp, "A Tool for Analyzing Voltage Collapse in Electric Power Systems," *Proceedings of the 10th Power Systems Computation Conference*, Graz, Austria, August, 1210-1217 (1990).

[25] A.M. Sasson, C. Treviño, F. Aboytes, "Improved Newton's Load Flow Through a Minimization Technique," *IEEE Trans. on Power Systems*, 90 (2), 1974-1981 (1971).

Hydrogenated Amorphous Silicon Thin films for Photovoltaic Technology:
Optical Probes of Growth Kinetics

MONICA KATIYAR

Department of Materials Science and Engineering
University of Illinois at Urbana-Champaign
Research Supervisor: Prof. J.R. Abelson

ABSTRACT

Thin film photovoltaic modules made of hydrogenated amorphous silicon (a-Si:H) will be a viable source of bulk electric power if they achieve a solar conversion efficiency of 15%. This requires improved quality of active materials, which are deposited on large-area substrates. Opto-electronic properties of a-Si:H are a function of its hydrogen content and microstructure. Here, in situ optical techniques, high sensitivity infrared spectroscopy to study hydrogen bonding and ellipsometry to study film microstructure, are used to relate the film properties to the controllable growth parameters. Results are presented for growth of a-Si:H on oxide substrates under various growth conditions. The hydrogen bonding in the first 10 Å of growth corresponds to the nucleation and coalescence stages of film growth, as confirmed by in situ ellipsometry. The experiments involving a-Si:H film growth on a-Si and isotope exchange provide the information about hydrogen incorporation and elimination rate at the growing surface. Further studies are required to relate these to film properties.

INTRODUCTION

Solar energy is an attractive alternative to the fast disappearing fossil fuel supply of the earth. It is a renewable source of energy and is environmentally safe without any toxic by-products. Although thin film photovoltaic (PV) modules directly converting solar energy into electric power have been successfully used in space applications for over three decades, there is a long way to go in terms of cost reduction for making this technology an economical source of bulk electric power. PV devices using thin films of hydrogenated amorphous silicon (a-Si:H) and its alloys can be produced cheaply over large areas and will be economically competitive for bulk electric power if their solar conversion efficiency can be improved to be ~15 %. Over the last decade, research programs for developing hydrogenated amorphous silicon and its alloys have been successful in reducing the cost of a-Si:H PV modules by five times. As a result, now large area modules are commercially available and field tested for medium scale power production. [1] Still further cost reduction is needed to gain the competitive edge over conventional sources of electric power. This can be attained by innovations in the design, materials and structure of the cells. In this report, I concentrate on the materials aspects.

In the first section, the basic operation of a-Si:H solar cells is discussed qualitatively to identify the problems that can be tackled by improving the quality of bulk materials and interfaces. Improving the material requires knowledge of correlation between growth parameters and film properties. Hydrogen content and microstructure of the a-Si:H films are empirically correlated with their electrical and optical properties. [2] Therefore, we focus on systematically studying the relationship between growth parameters and resulting hydrogen content and microstructure of the films. A technique sensitive to Si-H bonding is needed to understand the hydrogen incorporation in a-Si:H. Infrared spectroscopy is ideally suited to study the evolution of Si-H bonding during film growth: it distinguishes isolated monohydrides, which occur in the bulk amorphous network, from dihydrides and clustered Si-H bonds typical of surfaces. It can be used for real time, in situ probing of film growth in a plasma environment. In the second section, I describe this new reflection absorption infra red technique which is sensitive to ~5 Å thick a-Si:H film. The potential of this technique is evident from the studies of a-Si:H growth by reactive magnetron sputtering (RMS) on different substrates, oxide and a-Si, under various growth conditions. The evolution of Si-H bonding with thickness is related to the evolution of film microstructure observed by in situ ellipsometry, another optical probe sensitive to microstructure of the film. In view of the knowledge about growth flux in RMS form mass spectroscopy studies [3-5] one finds that both reactions at the surface by atomic hydrogen and implantation of energetic hydrogen below the surface play a role in deciding the final hydrogen content of the films. Future work is planned to correlate the opto-electronic properties of a-Si:H, including photo to dark conductivity ratio, midgap defect density, and bandgap, to hydrogen incorporation mechanism. This information is crucial to have control on material properties for making high efficiency cells.

SOLAR CELLS

Four types of device configurations have been used for solar cells in a-Si:H materials; homojunctions (p-i-n, p-n), Heterojunctions, Schottky Barrier, and metal insulator semiconductor structures.[6,7] Among all choices, the p-i-n structure as shown in figure 1(a) is found to be the most efficient, therefore the following analysis is mostly based on this structure. Figure 1(b) depicts the band diagram for the same structure. In this cell, photons are mainly absorbed by the i-layer, for the reasons discussed below, generating electrons and holes. If the generated carriers are within the diffusion length of the space charged region or in it, they get swept by the junction field to the opposite side giving rise to a photocurrent and photovoltage. In order to increase the conversion efficiency of solar cells one wants efficient absorption of solar energy to create carriers and then efficient collection of these carriers, before they recombine, to generate electricity. Due to the high defect density of p- and n-a-Si:H layers the photogenerated carriers recombine before they are swept by the junction. Therefore, in p-i-n junction as shown in figure 1(a) the thickness of the doped layer is only few 100 Å the minimum thickness required to sustain the space charge region to avoid any light absorption in this layer. On the other hand the intrinsic layer has lower defect densities and large mobility and lifetime products $(\mu\tau)$ for the carrier transport, so its thickness is chosen to create depletion region throughout the i-layer for better collection of photogenerated carriers.

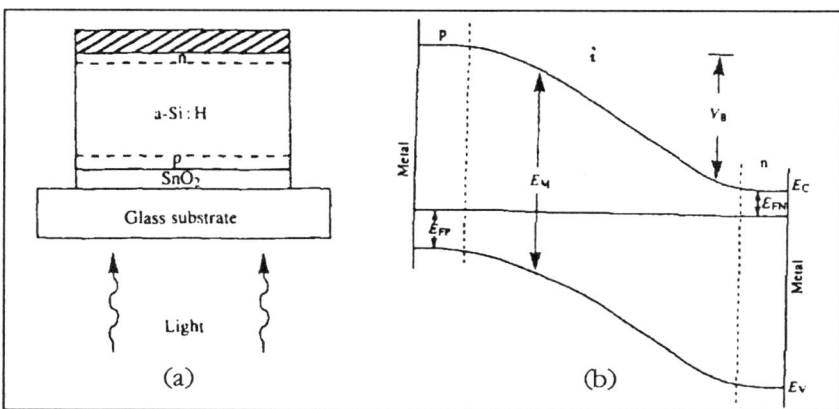

Figure 1 (a) Typical structure of single junction p-i-n solar cell. (b) Schematic diagram at zero bias showing the internal field of the undoped layer and the built-in potential V_B.

To quantify the efficiency of a solar cell one takes a p-i-n junction operating in the fourth quadrant of its I-V curve as shown in figure 2. [8] The conversion efficiency of the solar cell is given by

$$\eta = \frac{J_m V_m}{P_{in}} = \frac{J_{sc} V_{oc} FF}{P_{in}}, \tag{1}$$

where J_m and V_m are the output current density and voltage of a solar cell operating under maximum power conditions and P_{in} is the incident power density of the sunlight. J_{sc} and V_{oc} are short-circuit current density and open-circuit voltage as shown in the figure. FF is defined by the maximum area under a plot of photocurrent and photovoltage. Next we discuss each of these parameters to identify the material properties that affect conversion efficiency of solar cells.[8,9]

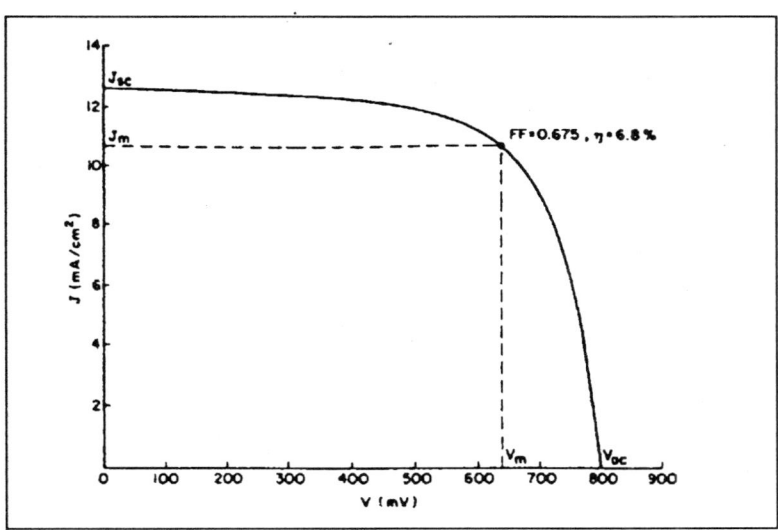

Figure 2. Current density as a function of voltage for an a-Si:H solar cell exposed to 100 mW cm^{-2} of simulated light. ref [8]

INCIDENT SOLAR ENERGY (P_{in})

The first step in converting solar energy to electric power is generation of photocarriers by absorption of sunlight. In order to increase the efficiency of this process the bandgap and thickness of the i-layer should be optimized for maximum absorption of the solar spectrum. The atmospheric solar spectra peaks at 5000 Å and drops rapidly below 4000 Å and extends far into the infrared. The calculations show that i-layer of ~1.7 eV bandgap will be optimum, but it does not absorb the low and high wavenumber region of the spectrum. To avoid these losses, stacked cells with three p-i-n junctions of low, normal and high bandgap i-layer have been proposed. For that purpose i-layers made of a-Si:H or its alloys are needed whose bandgap can be varied from 1.4 -2.0 eV. As for the doped layers, one needs minimum light absorption in these layers, therefore large bandgap materials like a-Si:C:H for p-layer and hydrogenated microcrystalline silicon (μc-Si:H) for n-layers need to be studied. Hydrogen plays an important role in deciding the bandgap of the a-Si:H and its alloys. It also plays an important role during growth e.g., in the deposition of μc-Si:H.

OPEN-CIRCUIT VOLTAGE (V_{OC})

Open circuit voltage can be increased in two ways.

1. By increasing the built-in potential of the device as shown in figure 1(b). This depends on the optical gap of the material, doping characteristics of the p- and n- layer and band bending at the p/i and i/n interface. Understanding the film growth will help in controlling some of these aspects of solar cell fabrication.
2. By minimizing the recombination of photogenerated carriers which takes place either in the i-layer or at the p/i or i/n interface. These centers could be either impurities or microstructure related.

SHORT-CIRCUIT CURRENT (I_{SC})

The short circuit current depends on the quality of the i-layer, i.e., large μt product and low defect density. Hydrogen passivates the defects that act as trapping, scattering and recombination centers, improving the mt product for the carriers.

FILL FACTOR (FF)

Fill factor depends on the series resistance of the device. From the materials point of view improving the conductivity of all layers will reduce the series resistance and increase FF.

FUNDAMENTAL GROWTH STUDIES

Optical bandgap of the layers, the μt product, doping efficiency and interface states that act as recombination centers, are some materials and growth problems that need to be addressed. As these properties are related to the hydrogen content of the films we want to understand the role of hydrogen during film growth and how it is incorporated in the films. a-Si:H films are deposited by plasma assisted processes e.g., glow discharge of silane and sputtering in Argon and Hydrogen plasma. In any plasma assisted deposition process the relationship between growth parameters and final properties is not a simple one. As shown in Figure 3, starting from the controllable growth parameters the first stage is understanding of plasma reactions that decide the growth flux coming on the substrate. The second stage is the reactions at the substrate that decide the final film properties. We focus our studies on a-Si:H film growth by reactive magnetron sputtering (RMS) which provides independent control of film hydrogen content via the hydrogen partial pressure (P_{H_2}) and substrate temperature (T_s). In addition, the RMS growth flux has been well characterized as described below:[3-5]

1. *Atomic Hydrogen*: A great fraction of atomic hydrogen flux is energetic hydrogen that are reflected from the target surface and possess average energy of ~200 eV with a broad energy distribution.
2. *H ions*: Hydrogen also arrives at the growing surface via the ions ArH^+, H_2^+, H_3^+ due to plasma reactions.
3. *SiH and silane radicals*: The SiH are sputtered from the silicon target and silane radicals are produced in various plasma reactions.

Having some knowledge of growth flux, I now describe the optical diagnostics used to study the reactions at the growth surface.

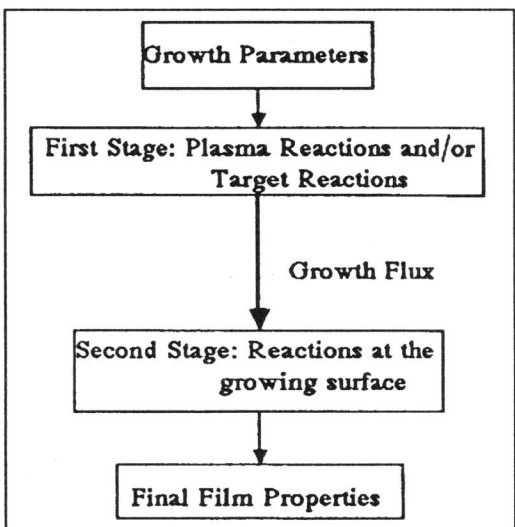

Figure 3. Typical flow chart for plasma assisted deposition of the materials.

REFLECTION ABSORPTION INFRARED SPECTROSCOPY (RAIRS)

Generally the IR absorption due to vibrational modes is weak and, therefore, not sensitive to very thin films or surfaces. In order to study surface reactions we require an enhancement technique suitable for practical thin film growth. We have developed multilayer substrates to enhance IR absorption in a reflection mode and detect Si-H modes in films only 5 Å thick. As shown in Figure 4, the substrates are 300-350 μm thick double polished c-Si wafers with 8500 Å thick thermal oxide grown on both sides. The back surface is coated with aluminium and a-Si:H is deposited on top. We use a computer program which rigorously incorporates the effects of coherent multiple reflections in the a-Si:H and SiO_2 layers and incoherent reflections in the c-Si to quantitatively relate the sample reflectance to the a-Si:H optical constants. These optical constants in turn are converted to hydrogen concentration using known oscillator strength. These substrates provide ~15 times larger IR absorption than in transmission measurements of a-Si:H on c-Si. In addition, real time measurements require that this enhancement must not be very sensitive to film thickness and morphology for interpretation. On the multilayer substrates the bulk and near-surface regions are probed with equal sensitivity for ≤ 400 Å of a-Si:H deposition. In contrast, for films on metal substrates, the H-rich surface layer is probed much more preferentially than the bulk. [10] If we represent the absorption in terms of E_2, the imaginary part of dielectric constant, then enhancement does not depend on E_1, the real part of the film dielectric constant. This allows us to use the same enhancement factor for all thicknesses and for different films (different E_1 values). Another advantage of these substrates is constant background spectrum that simplifies the analysis. Because of this we are able to use the reflectance from bare substrate as the reference.

The experimental set-up is simple: a 2" diameter mini magnetron source with a high purity c-Si target is sputtered in d.c (Ar-H_2) plasma to deposit the a-Si:H films. Argon pressure is kept constant at 1.6 mTorr and hydrogen partial pressure (P_{H_2}) and substrate Temperature (Ts) are varied to change the hydrogen content of the films from 0-30 atom%. A schematic of the deposition chamber is shown in Figure 5. The infrared beam from an Analect FTIR spectrometer (RFX-65) is focused onto the growing film through a KBr window. The angle of incidence is 70°. The beam reflected from the sample is focused onto a high sensitivity HgCdTe detector. All the spectra are collected with the s-polarization. Spectra are recorded over the frequency range 400-4800 cm^{-1} by averaging 128 scans, which takes 25 seconds. The films are deposited at low deposition rate to get better thickness resolution. Normally spectra are taken every 2-4 Å of film growth depending on the growth conditions. We focus on stretching modes of Si-H, 1800-2500 cm^{-1} for two reasons:

1. HgCdTe detectors are not very sensitive in low wavenumber region, making it difficult to see the SiH_2 bending modes (800-900 cm^{-1}) and the SiH wagging mode (640 cm^{-1}).

2. The stretching modes give the information about the bonding configuration of the hydrogen: The peak at 2000 cm^{-1} is attributed to the isolated monohydride Si-H bond in the bulk. The peak at 2100 cm^{-1} is due to the dihydride bonds and also due to Si-H clusters on internal voids. In addition, we find a narrow component at 2100 which we attribute to all SiH.

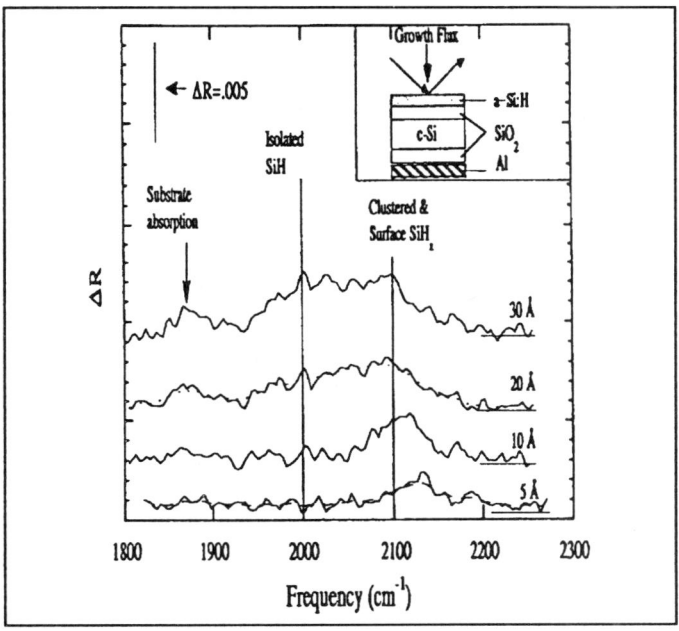

Figure 4.   IR absorption spectrum at different time intervals during deposition. The film is grown at P_{H_2}=0.1 mTorr. Spectra are plotted with offset for clarity. Dashed lines are drawn just to guide the eye. Insert shows the schematic of the substrate geometry.

Details of the data collection are described in our previous papers.[11,12] The major drawback of IR spectroscopy is that it does not give any explicit information about film microstructure and morphology. Therefore, results from in situ ellipsometry done on the same system, as shown in Figure 5, are used to get this information.

RESULTS AND DISCUSSION

Figure 4 shows four infrared spectra taken in real time during the initial stages of a-Si:H growth with P_{H_2} = 0.1 mTorr. Here, DR = R_S-R, where R is the sample reflectance and R_S is the bare substrate reflectance. In the first 5-10 Å of growth only a narrow peak (FWHM=20 cm^{-1}) centered at 2120 cm^{-1} is visible. For larger thicknesses (d) this peak broadens (FWHM=40 cm^{-1}) and shifts down to 2100 cm^{-1} and a peak at 2000 cm^{-1} appears. This trend is qualitatively similar to the one seen in our previous ex situ measurements.[13] IR studies of H on c-Si reveal that *all* Si-H bonds, i.e., mono-, di-, and trihydride species appear at ~2070-2140 cm^{-1}.[14] We therefore attribute the narrow peak at 2120 to the stretching modes of all SiH$_x$ on the surface of the growing a-Si:H film. The 2000 cm^{-1} mode has been attributed to isolated monohydrides in a-Si:H. We observe the 2000 cm^{-1} mode when d is greater than 10-15 Å and interpret this as the thickness necessary to form a bulk like film. This is significantly lower than the value of 100 Å for rf glow discharge deposited a-Si:H reported by Toyoshima et al.[15] and similar to the value reported by Blayo and Drevillon.[16] We believe that the high value obtained in reference [15] is due to the low sensitivity of bulk hydrogen for metal substrates as discussed above.

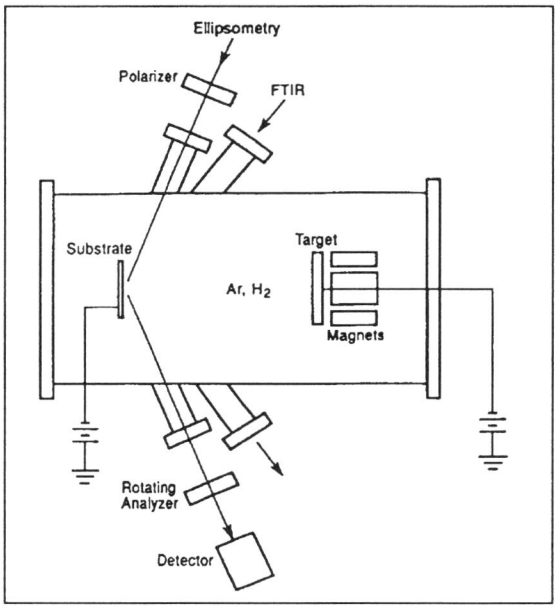

Figure 5. Schematic of the growth chamber with the in situ diagnostics.

Since the IR beam probes the entire depth of the sample, DR integrated over an absorption band is proportional to the areal density of Si-H bonds, and the slope of DR vs. d is proportional to the rate of net H incorporation. The proportionality factors are the oscillator strengths 1.8x10^{20} cm^{-2} for the 2000 cm^{-1} mode and 4.4x10^{20} cm^{-2} for the 2100 cm^{-1} modes, [17] obtained from transmission IR measurements on thick films, and the enhancement factor obtained from the simulations.[10] Using these values Figure 6 shows the areal density of Si-H bonds in the 2100 cm^{-1} mode (surface SiH$_x$, clustered SiH and SiH$_2$) vs. film thickness for P_{H_2} = 0.1, 0.3, 0.8 and 1.0 mTorr and T$_s$ = 200°C. For all films the evolution can be divided into three regions. For d increasing from 0-10 Å, the 2100 cm^{-1} mode increases rapidly and linearly; within experimental resolution the spectra are independent of P_{H2}. For d between 10 and 25 Å, the rate of increase is considerably smaller than in the first 10 Å and is weakly dependent on P_{H_2}. For d>25 Å, the 2100 cm^{-1} mode increases linearly, the rate (slope) being higher for higher P_{H_2}. We also did these measurements as a function of T$_s$ and found very similar trends in the aerial density of 2100 cm^{-1} mode versus d, the steady state slope (d>25 Å) is smaller for high T$_s$. These results will be discussed later from the point of morphology development as determined by in situ ellipsometry.

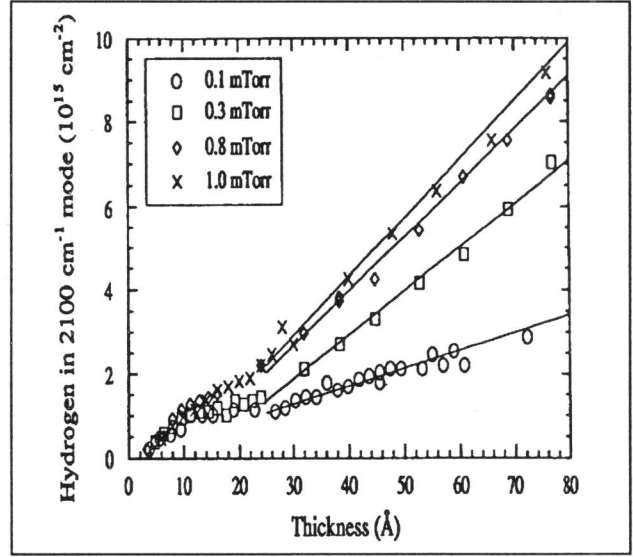

Figure 6.   Increase in the hydrogen content of 2100 cm^{-1} mode with film thickness at different hydrogen partial pressures (0.1 to 1.0 mTorr).

Figure 7 shows the hydrogen in 2000 cm^{-1} mode absorption vs. film thickness for the films of Figure 6. For all deposition conditions this mode only appears for d >15 Å and increases linearly with thickness. This is due to the growth of bulk like material at this point, although from trends in 2100 cm^{-1} the growth is still not in steady state until > 25 Å. Note that the hydrogen content in the 2000 cm^{-1} mode increases from P_{H_2}=0.1 to 0.3 mTorr and decreases for 0.8 mTorr. This trend is consistent with the transmission measurements on thick films.

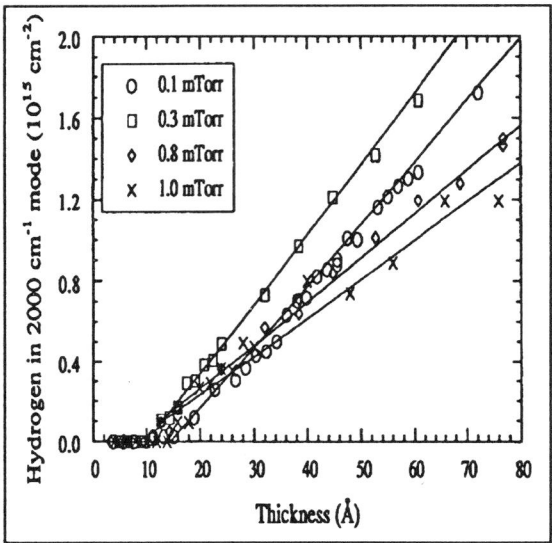

Figure 7. Increase in the hydrogen content of 2000 cm^{-1} mode with film thickness for same conditions as in Figure 6.

Figure 8 shows the total hydrogen content, i.e. both 2000 and 2100 cm^{-1} modes, for the same films shown in Figures 7 & 6. The initial 10 Å of film is rich in hydrogen content compared to the rest of the film, i.e. the slope for d <10 Å is larger than the slope for d >25 Å. The areal density of hydrogen, H=1.2±0.2 x10^{15} cm^{-2} at d=10 Å, is the same for all growth conditions. The hydrogen incorporation rate is small for the next 10-15 Å of growth and then increases for d >25 Å to the steady state bulk incorporation rate. The evolution of IR spectra is consistent with the evolution of microstructure determined by multiple wavelength ellipsometry results for film growth on c-Si substrates with native oxide.[12] The latter shows that a-Si:H growth proceeds via the nucleation of islands. Coalescence occurs when the average island radius increases to ~10-13 Å which is a mass thickness of 5 Å (we cite the latter). Beyond this the film grows layer by layer with a small decrease in surface roughness. The IR signal is only in 2100 mode up to 10 Å of growth. Whether the 2100 cm^{-1} mode observed for d< 10 Å is due to this surface layer or a H-rich interface with the substrate still needs to be clarified. Within the experimental uncertainty we do not see any changes in the first 10 Å of growth with P_{H_2} and T_s similar to in situ ellipsometry results that show no significant change in maximum island radius of nuclei. Table I compares the total hydrogen content obtained from reflectance measurements (slope for d> 25 Å) with values from transmission measurements on thick films grown on c-Si under the same conditions. The two values agree remarkably well, suggesting that bulk-like Si-H bonding is established in >15

Å thick films. This is consistent with ellipsometric measurements which show that film microstructure is independent of thickness for d > 10-15 Å.[12]

Table I. Deposition conditions and hydrogen content of a-Si:H films determined from absorption at the stretching modes (2000 and 2100 cm-1) for reflectance measurements and from wagging mode (640 cm-1) for transmission measurements.

P_{H2} (mTorr)	$T_{substrate}$ (°C)	C_H 2000 cm-1 (atom %)	C_H 2100 cm-1 (atom %)	$C_{H,total}$ reflectance (atom %)	$C_{H,total}$ transmission (atom %)
0.1	200	4.8	8.0	12.8	11.7
0.3	200	6.0	18.3	24.3	22.2
0.8	200	3.8	22.5	26.3	24.4
1.0	200	3.1	24.2	27.3	-------
0.3	120	10.2	23.0	33.2	30.4
0.3	270	6.1	7.5	13.6	14

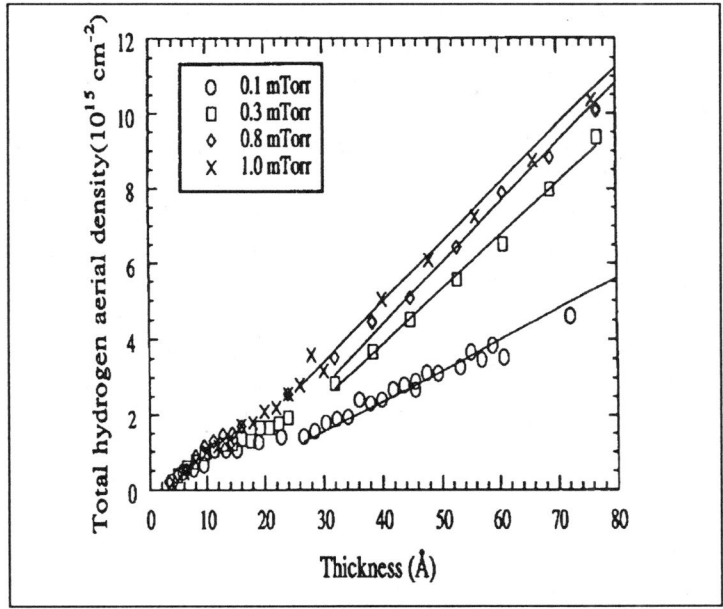

Figure 8.   Total hydrogen content of the film as a function of film thickness (d) at different hydrogen partial pressures (0.1 to 1.0 mTorr).

For film thicknesses of 10-25 Å, however, the 2100 cm-1 mode is not growing at the steady state rate until d >25 Å. In this region the slope increases slightly with increasing H partial pressure, and decreases with increasing substrate temperature (data not shown). There are three possibilities for the existence of plateau region: i) the film growing near the interface is very different in H bonding compared to the steady state bulk film (although there is no microstructural difference evident in ellipsometry data); ii) the surface roughness decreases as the thickness increases, as seen by in situ ellipsometry, [12] so the decrease in 2100 mode is due to decrease in surface area which compensates the increase in the 2100 mode from the bulk growth; iii) H is implanted 20-25 Å below the surface due to the energetic H flux present in magnetron sputtering and this fraction of hydrogen will not show up in the film until d >25 Å. Hence, after first 10 Å of nucleation and coalescence stage the hydrogen incorporation rate is lower than the steady state slope of d>25 Å due to the absence of hydrogen incorporation in the film by implantation. Due to the transient nature of growth in the first 25 Å the data is not easily interpretable.

To isolate surface and subsurface contributions, compare the growth of a-Si:H, as shown in Figure 9, on the oxide substrate and on a-Si that was deposited on the multilayer substrate. The rapid rise of 2000 cm^{-1} mode, attributed to Si-H bonds in the bulk, for the growth on a-Si film indicates that some of the hydrogen is getting implanted in the a-Si. The offset in the curves points to the hydrogen incorporation in the underneath a-Si layer. The question is whether it is due to diffusion or implantation. This experiment gives information about the kinetics of hydrogen incorporation, but it is not in steady state growth. Therefore, another experiment is done involving isotope switching between hydrogen and deuterium: when D_2 is injected into the plasma instead of H_2, both the surface and bulk H modes decrease as shown in figure 10, supporting the subsurface reaction hypothesis. Assuming growth kinetics does not change significantly upon switching dueterium for hydrogen, from Figure 10, we can obtain elimination rate of hydrogen from the growing surface under steady state growth conditions. Experiments with the growth of a-Si:H on unhydrogenated and deuterated a-Si are underway to quantify the hydrogen incorporation and elimination rates at the growing surface.

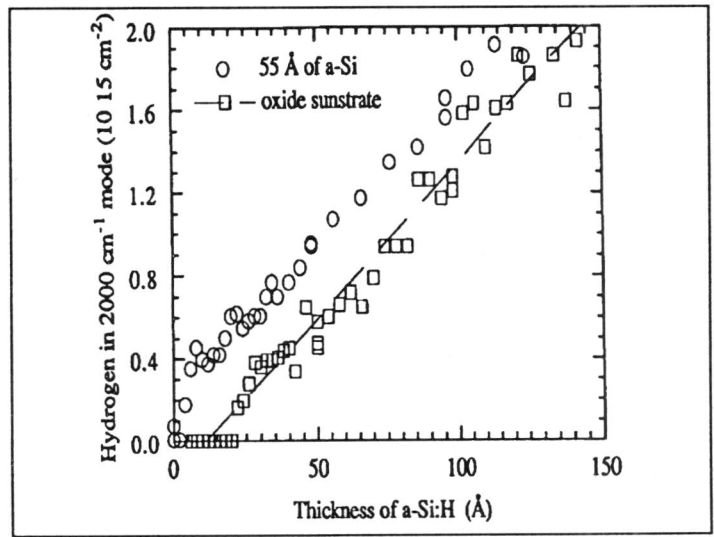

Figure 9. Comparison of 2000 cm^{-1} mode for a-Si:H growth on 55 Å of a-Si and oxide showing the rapid rise of the mode for growth on a-Si. Deposition conditions are P_{H_2}=0.3 mTorr, T_{sub}=230 °C.

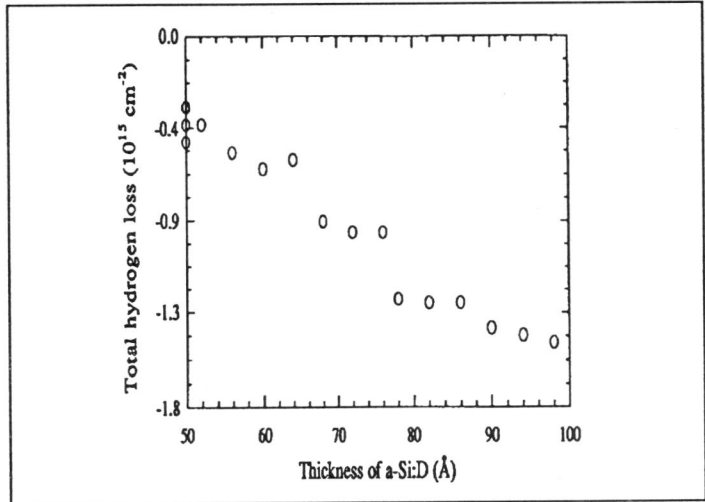

Figure 10. Hydrogen loss upon growth of a-Si:D on 55 Å of a-Si:H indicating hydrogen elimination processes during steady state film growth, i.e., away from nucleation stage. Deposition conditions are P_{H_2}/D2=0.3 mTorr, T_{sub}=230 °C.

CONCLUSIONS

In summary, we use a multilayer optical cavity substrate to enhance infra-red absorption, and monitor the real time evolution of Si-H bonding during the growth of a-Si:H. Only the 2100 cm^{-1} mode is detected in the first 10 Å of film growth, in agreement with nucleation and coalescence stages. The initial incorporation is independent of growth conditions and points to a hydrogen rich surface or interface layer. Non uniformity of hydrogen concentration profile over next 20 Å indicates complex H incorporation and release mechanisms. Quantitative estimation of hydrogen content from in situ measurements is in good agreement with the standard ex situ measurements.

From the results of a-Si:H growth on a-Si and isotope exchange experiment we believe that the hydrogen concentration of the film is decided partly by surface reactions and partly by diffusion and implantation. Quantitative estimation of these reactions will be carried out in the proposed research. The results to date have shown that real time, in situ IR reflectance spectroscopy during growth has both the sensitivity and quantitative interpretation needed to study surface and bulk processes involving hydrogen in a plasma environment.

FUTURE WORK

These results are very encouraging, but this work is still in a preliminary stage. The following experiments are planned to understand the growth kinetics quntitatively and relate it to film properties:

1. *Growth of a-Si:H on a-Si*: Develop a model to explain the evolution of hydrogen bonding for the growth of a-Si:H on a-Si films in terms of implantation and diffusion effects with physically reasonable parameters. These include estimates of the kinetic energy distribution and flux of the energetic hydrogen, and diffusion coefficient of H in solids. TRIM will be used to get an estimate of the implantation depth.
2. *Growth of a-Si:D on a-Si:H*: We will look at the hydrogen elimination kinetics doing isotope exchange studies, discussed above as a function of thickness of the a-Si:H layer, D_2 partial pressure and substrate temperature.
3. Correlations will be sought between the opto-electronic properties of a-Si:H, including photo to dark conductivity ratio, midgap defect density, bandgap, and hydrogen incorporation under widely varying conditions.
4. Similar isotope switching measurements will be done under conditions which produce mc-Si:H films.

Acknowledgments: Funding from Link Energy Foundation and Thin Film Solar Cell Program of the Electric Power Research Institute under Contract # EPRI RP 8001-7 is gratefully acknowledged. I wish to thank Dr. G.F. Feng, Dr. N. Maley and Prof. J.R. Abelson for their help and guidance.

REFERENCES

[1]   T. Peterson, Amorphous Silicon Thin-film Photovoltaics, EPRI Journal, 44-47 (June 1991).

[2]   Properties of Amorphous Silicon, INSPEC, The Institution of Electrical Engineers, London and New York, 129-461 (1989).

[3]   J.R. Abelson, N. Maley, J.R. Doyle, G.F. Feng, M. Fitzner, M. Katiyar, L. Mandrell, A.M. Myers, A. Nuruddin, D.N. Ruzic, and S. Yang, In situ Measurements of Hydrogen Flux, Surface Coverage, Incorporation and Desorption during Magnetron Reactive Sputtering, Mat. Res. Soc. Symp. Proc., *219.*, 619-630 (1991).

[4]   A. Myers, Characterization of the Growth Flux during the Deposition of Hydrogenated Amorphous Silicon by DC Magnetron Reactive Sputtering, Ph.D Thesis, University of Illinois, 131-252 (1991).

[5]   J.R. Abelson, J.R. Doyle, L. Mandrell, A.M. Myers, and N. Maley, Surface Hydrogen Release during the Growth of a-Si:H by Reactive Magnetron Sputtering, J. Vac. Sci. Technol. A, *8(3)*,1364-1368 (1990).

[6]   A. Madan, and M.P. Shaw, Opto-Electronic Applications of Amorphous Silicon Based Materials, in The Physics and Applications of Amorphous Semiconductors, Academic Press, Inc., San Diego, 163-275 (1988).

[7]   D.E. Carlson, and C.R. Wronski, Amorphous Silicon Solar Cells, in Amorphous Semiconductors, Edited by M.H. Brodsky, Springer-Verlag, Berlin, Heidelberg, 287-330 (1985).

[8]   D.E. Carlson, Solar Energy Conversion, in The Physics of Hydrogenated Amorphous Silicon I: Structure, Properties, and Devices, Edited by J.D. Joannopoulos and G. Lucovsky, Springer-Verlag, Berlin, Heildelberg, 203-244 (1984).

[9]   R.R. Arya, High Efficiency Amorphous Silicon Based Solar Cells: A Review, Mat. Res. Soc. Sym. Proc., *118*, 569-580 (1988).

[10]   N. Maley and I. Szafranek, to be published.

[11]   M. Katiyar, G.F. Feng, J.R. Abelson and N. Maley, In Situ IR Absorption Study of H Bonding in a-Si:H Thin films, Mat. Res. Soc. Symp. Proc., *219*, 295-300 (1991).

[12]   G.F. Feng, M. Katiyar, J.R. Abelson and N. Maley, Substrate Induced Crystallinity in Reactive Sputter Deposition of Hydrogenated Silicon, Mat. Res. Soc. Symp. Proc., *219*, 709-714 (1991).

[13]   N. Maley, I. Szafranek, L. Mandrell, M. Katiyar, J.R. Abelson and J.A. Thornton, Infrared Reflectance Spectroscopy of Very Thin Films of a-SiH, J. Non-Cryst. Sol., *114*, 163-165 (1989).

[14]   Y.J. Chabal, G.S. Higashi, K. Raghavachari, V.A. Burrows, Infrared Spectroscopy of Si(111) and Si(100) Surfaces after HF Treatment: Hydrogen Termination and Surface Morphology, J. Vac. Sci. Technol. A, 7(3), 2104-2109 (1989).

Parallel Arrays of Josephson Junctions for Submillimeter Local Oscillators[*]

ALEKSANDAR PANCE
and Michael J. Wengler

Department of Electrical Engineering
University of Rochester
Rochester, NY 14627

[*] Also in Proceedings of the Third International Symposium on Space Terahertz Technology, University of Michigan, Ann Arbor, Michigan, March 24-26, 1992.

ABSTRACT

In this paper we discuss the influence of the DC biasing circuit on operation of parallel biased quasioptical Josephson junction oscillator arrays. Because of nonuniform distribution of the DC biasing current along the length of the bias lines, there is a nonuniform distribution of magnetic flux in superconducting loops connecting every two junctions of the array. These DC self-field effects determine the states of the array. We present analysis and time-domain numerical simulations of these states for four biasing configurations. We find conditions for the in-phase states with maximum power output. We compare arrays with small and large inductances and determine the low inductance limit for nearly-in-phase array operation. We show how arrays can be steered in H-plane using the externally applied DC magnetic field.

INTRODUCTION

The Josephson junction is a natural choice for submillimeter local oscillator since it is a "voltage controlled oscillator" with typical voltage scales of mV and an oscillation frequency f_J = 483 GHz per mV of dc bias. The existence of Josephson radiation into the terahertz range has been demonstrated at Cornell [1]. A major disadvantage of the Josephson junction is its very low output power. With DC voltage bias of 1 mV at 483 GHz, a junction which could accept 100 μA will put out less than 100 nW of RF power. Therefore, practical local oscillators must use arrays of many junctions oscillating in phase. Submillimeter Josephson oscillator arrays with usable power levels have been made at Stony Brook [2] and NIST [3].

We have proposed to build a large 2-D active grid array of parallel biased Josephson junctions [4]. In our design, every junction drives a single antenna and the power from the whole array is quasioptically combined. By biasing all junctions in parallel, we assure that all of them radiate at exactly the same frequency. For maximum radiated power, all junctions must also be in phase.

The DC biasing circuit of the 2-D quasioptical Josephson array plays a very important role in phase-locking of Josephson junctions. In a two-dimensional array the DC biasing current is supplied at the ends (Fig. 1). Because of that, the DC current is nonuniformly distributed along the length of the biasing line. This current induces the nonuniform DC magnetic flux in superconducting loops between every two neighboring junctions. Because of the superconducting quantum interference effects [5], these self-induced fluxes determine the phase differences between the neighboring junctions, and therefore the states of the array. These effects will be referred to as the DC self-field effects. It is clear that, depending on the particular bias circuit, the in-phase state can only be a special, rather than common state of the parallel 2-D Josephson junction array.

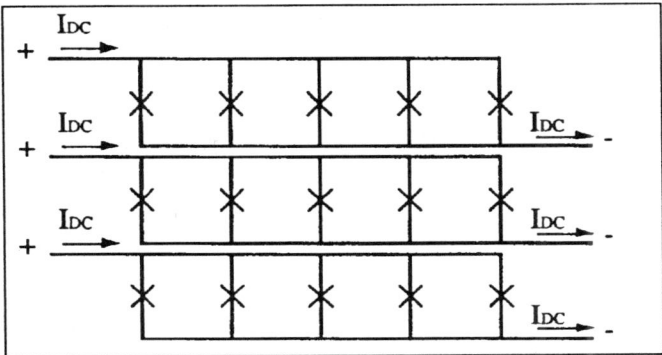

Figure 1. Separate row bias for parallel 2-D Josephson junction array.

If the rows of the 2-D parallel Josephson array are biased independent from each other, the DC self-field effects are, to the first order, limited to each row, and the whole 2-D array can be looked at as a collection of 1-D parallel arrays. We will therefore investigate these DC self-field effects in linear parallel arrays.

N-JUNCTION LINEAR PARALLEL ARRAY

The most general biasing scheme for the linear parallel array is presented in Fig. 2. We use the RSJC model of the Josephson junction that consists of ideal Josephson junction, shunt resistance and parasitic capacitance (Fig. 2). The ideal Josephson junction is described by relationships between its current I, voltage V and phase difference ϕ of the superconducting quantum mechanical wave function between two sides of the junction

$$I = Ic \sin (\phi), \frac{d\phi}{dt} = \frac{2e}{\hbar} V,$$

where Ic is the junction critical current. Assuming that all junctions are identical, the circuit from Fig. 2 can be described with the following system of equations:

$$i_1(\tau) = \frac{1}{\lambda} (\phi_2(\tau) - \phi_1(\tau)) + \frac{\gamma_1}{2} - \frac{\gamma_R}{2} - \frac{2\pi}{\lambda} \varphi_{ex},$$

$$i_j(\tau) = \frac{1}{\lambda} (\phi_{j+1}(\tau) - 2\phi_j(\tau) + \phi_{j-1}(\tau)) + \frac{\gamma_j}{2}, \quad j = (2, N-1),$$

$$i_N(\tau) = -\frac{1}{\lambda} (\phi_N(\tau) - \phi_{N-1}(\tau)) + \frac{\gamma_N}{2} + \frac{\gamma_L}{2} + \frac{2\pi}{\lambda} \varphi_{ex}, \tag{1.a}$$

where ϕ_j is the superconducting wave function phase difference across the jth junction, i_j is the total current through the jth junction, γ_j, γ_L and γ_R are biasing currents and φ_{ex} is the normalized externally applied DC magnetic flux,

$$i_j(\tau) = \beta \ddot{\phi}(\tau) + \dot{\phi}(\tau) + \sin (\phi(\tau)),$$

$$\gamma_j = i_{Uj} + i_{Dj},$$

$$\gamma_L = i_{Lin} + i_{Lout}, \quad \gamma_R = i_{Rin} + i_{Rout},$$

$$\varphi_{ex} = \frac{\Phi_{ex}}{\Phi_0} \qquad (1.b)$$

with capacitance and inductance parameters β and λ, respectively, given by

$$\beta = \frac{R^2 C}{L_j}, \quad \lambda = \frac{L}{L_j}, \quad L_j = \frac{\Phi_0}{2\pi I_c}, \quad \Phi_0 = \frac{h}{2e}, \qquad (1.c)$$

where L_j is the zero-bias Josephson junction inductance and $\Phi_0 = 2.07 \times 10^{-15}$ Wb is the flux quantum. In equations (1.a-c), time is given in units of $\frac{L_j}{R}$, all currents in units of I_c and normalized junction voltages, that are just time derivatives of junction phases ϕ_j, in units of $I_c R$.

In the case of 2-junction array, the in-phase solution has been reported by Ben-Jakob et al [6]. Here we present solutions for several N-junction arrays with different biasing configurations: the LL ("left"-"left"), LR ("left"-"right"), UD ("up"-"down") and CB ("central bias") biased array (Fig. 3). Although the LL and LR bias are directly applicable to parallel biased two-dimensional arrays, the other two configurations are used in other array architectures, such as series-parallel combinations, etc.

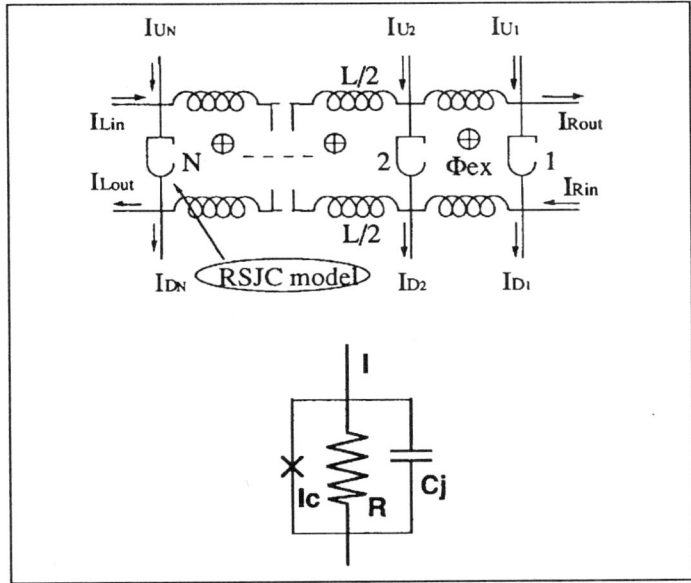

Figure 2. General biasing scheme for one-dimensional parallel Josephson junction array. The RSJC model used is shown below.

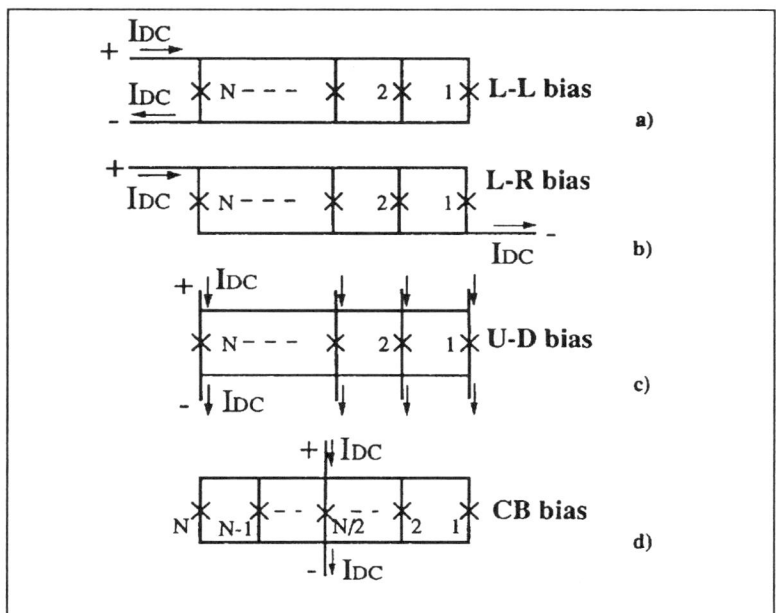

Figure 3.   Four common biasing configurations of 1-D array.

IN-PHASE STATES

The general solution of eq. (1) for the junction phases ϕ_j is

$$\phi_j(\tau) = \omega\tau + f_j(\tau) + \phi_j(0),$$

$$\omega = \langle \dot{\phi}_j(\tau) \rangle,$$

$$f_j(\tau + T) = f_j(\tau), \quad T = \frac{2\pi}{\omega}, \quad \langle f_j(\tau) \rangle = 0, \tag{2}$$

where "$< >$" denotes time average, ω is the normalized DC voltage across junctions, f_j's are some general, periodic functions with zero time-average and $\phi_j(0)$ are constants. For the in-phase solution, the following condition must hold for every two neighboring junctions

$$\phi_{j+1}(\tau) - \phi_j(\tau) = 2\pi m_j, \tag{3}$$

where m_j must be an integer. Note that m_j represents the number of fluxons in the jth loop.

Condition (3) is fulfilled if

$$f_j(\tau) = f(\tau),$$

$$\phi_{j+1}(\tau) = \phi(\tau + \tau_j),$$

$$\tau_j = m_j T, \tag{4}$$

where τ_j is the time delay between the phases of the two neighboring junctions. Substituting eq. (3) into the system (eq. 1.a) and equating all the currents i_j leads to solutions

$$m_j = j\,a + \varphi_{ex}, \qquad\qquad \text{(LL bias)} \qquad \text{(5.a)}$$

$$m_j = (j - \frac{N}{2})\,a + \varphi_{ex}, \qquad \text{(LR bias)} \qquad \text{(5.b)}$$

$$m_j = \varphi_{ex}, \qquad\qquad\qquad \text{(UD bias)} \qquad \text{(5.c)}$$

$$m_j = ([\frac{N}{2}] - j)\,a + \varphi_{ex}, \qquad \text{(CB bias)} \qquad \text{(5.d)}$$

where [ ] in eq. 5.d denotes integer division, and parameter 'a' is given by

$$a = \frac{\lambda i_{DC}}{2\pi N}, \tag{6}$$

where i_{DC} is the total DC biasing current. In order for m_j to be an integer, which is the precondition for the in-phase solution, it is necessary that both φ_{ex} and a be integers:

$$\varphi_{ex} = k_\varphi,$$

$$a = \frac{\lambda i_{DC}}{2\pi N} = k, \tag{7}$$

where k_φ and k are integers. The only exception is the LR array with odd number of junctions N, where "a" must be an even integer (2 k). The arrays will be in phase for all currents i_k that satisfy

$$i_k = k\frac{2\pi N}{\lambda}. \tag{8}$$

Note that these in-phase states are achieved without external locking mechanisms.

NUMERICAL SIMULATIONS

System (1) is solved numerically using the 4th order Runge-Kutta method [7]. Figure 4 shows the I-V and dV/dI-I curves of the 4 junction LL biased array with $\lambda=20$ and $\beta=0.5$. According to eq. (8), the in-phase states appear for current bias $i_k= 1.256$ k ($i_k'=0.314$ k for bias current normalized to the number of junctions, N, as in the Figure 4). The in-phase states are visible as voltage maximums in the I-V curve and sharp and deep minimums in dV/dI-I curve, for k=4 to 7. Similar structure has been observed experimentally by Clarke et al [8].

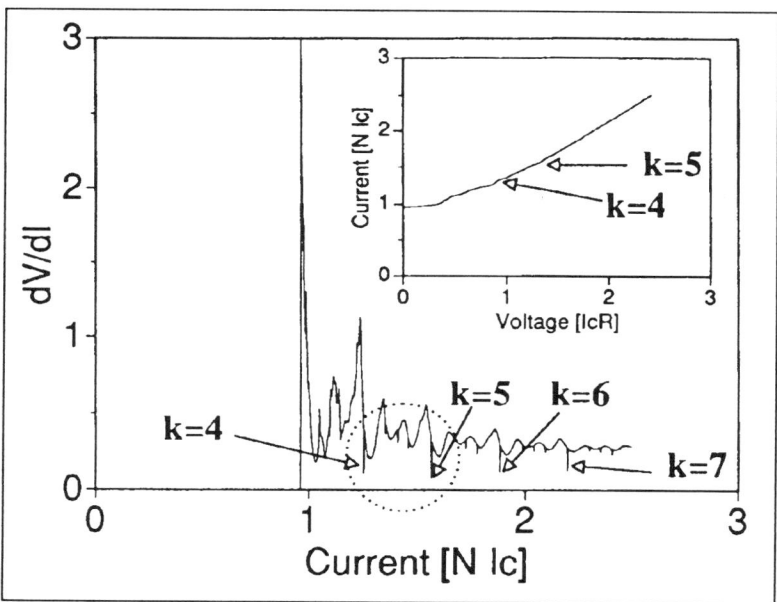

Figure 4. Dynamic resistance and I-V curve of 4 junction LL biased array with $\lambda=20$ and $\beta=0.5$. The in-phase states are seen as small steps in the I-V curve and sharp minimums in dV/dI-I curve, labeled k=4 to k=7. The area inside a circle is shown enlarged in Fig. 5.

OTHER STATES

The dV/dI-I curve of Figure 4 reveals considerable periodic structure between the in-phase states. Under certain conditions, that will be specified below, these "other" states, for current bias $i_{DC} \neq i_k$, correspond to the general solution of eq. (2) that satisfies the following:

$$f_{j+1}(\tau) - f_j(\tau) = \varepsilon_j(\tau),$$

$$\phi_{j+1}(\tau) \approx \phi_j(\tau + \tau_j),$$

$$\tau_j \approx \mu_j T, \qquad\qquad\qquad (9)$$

where $\varepsilon_j(\tau)$ is an error term and μ_j does not need to be an integer. Furthermore μ_j is found from the same equations as m_j (eq. 5), except that a and φ_{ex} are no more restricted to integer numbers. In other words, all states of the parallel array are described with phases at neighboring junctions shifted in time by an amount determined by the DC biasing current and external magnetic field (eq. 5). It is convenient to define the relative normalized time shift θ_j between the waveforms of functions f_{j+1} and f_j

$$\theta_j = \tau_j \bmod T = \mu_j \bmod 1, \qquad\qquad (10)$$

where "mod" is the modulus function, so that $0 \leq \theta_j \leq 1$. It has been shown by perturbation analysis [6] that in the case of 2-junction array solution, eq. (9), holds in the neighborhood of the in-phase state $(i_{DC} = i_k + \Delta i_{DC})$ and it has been suggested that it will hold for any state between the in-phase states, for the case of weak coupling $(\lambda >> 1)$.

Figure 5.a shows the circled part of the dV/dI-I curve of Fig. 4. Points labeled "1" and "4" correspond to the in-phase states with k=4 and k=5, respectively. The voltage waveforms on individual junctions for these two states are shown in Figures 6.a and 7.a, respectively. Point "2" of Fig. 5.a correspond to DC biasing current $i_{DC} = 5.65$, so eq. (5.a) gives $\mu_1 = 4.5$, $\mu_2 = 5$ and $\mu_3 = 5.5$ for the number of fluxons in each loop. From eq. (10) we find that relative time shifts should be $\theta_1 = 0.5$ between the voltages of the junctions 2 and 1, $\theta_2 = 0$ for junctions 3 and 2 and $\theta_3 = 0.5$ between junctions 4 and 3. Numerical simulations shown in Figure 6.b confirm this prediction.

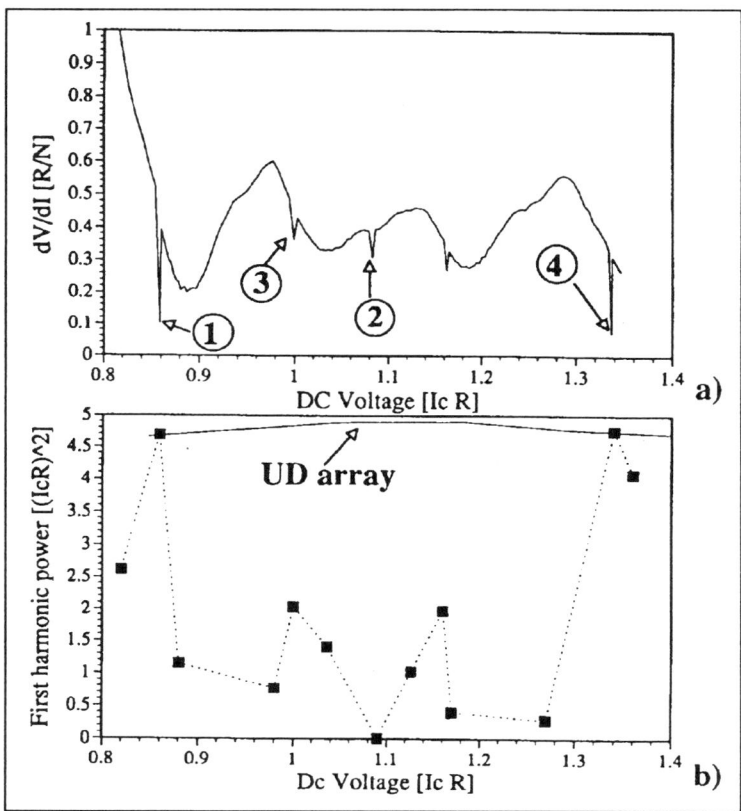

Figure 5.   Four junction array, LL bias, λ=20, β=0.5.
a) Enlarged portion of the dV/dI-I curve (Fig. 4) with in-phase
states labeled "1" and "4" and two "other" states "2" and "3".
The waveforms of individual junction voltages for these states
are shown in Fig. 6 and 7. b) First harmonic power that
correspond to states in a). The power is maximum in the
in-phase states and equal to that of the equivalent UD array.

Point "3" of Figure 5.a correspond to i_{DC} = 5.42, and again from equations (5.a) and (10) we obtain θ_1=0.333, θ_2=0.666 and θ_3=0. The voltage
(and phase) at junction 2 is time shifted by third of a period from that of
junction 1, voltage at junction 3 is shifted by two-thirds from that of junction
2, so that it is in phase with junction 1. Finally, junction 4 is in phase with
junctions 1 and 3. This situation is shown in Figure 6.c. All other states can
be determined in a similar fashion.

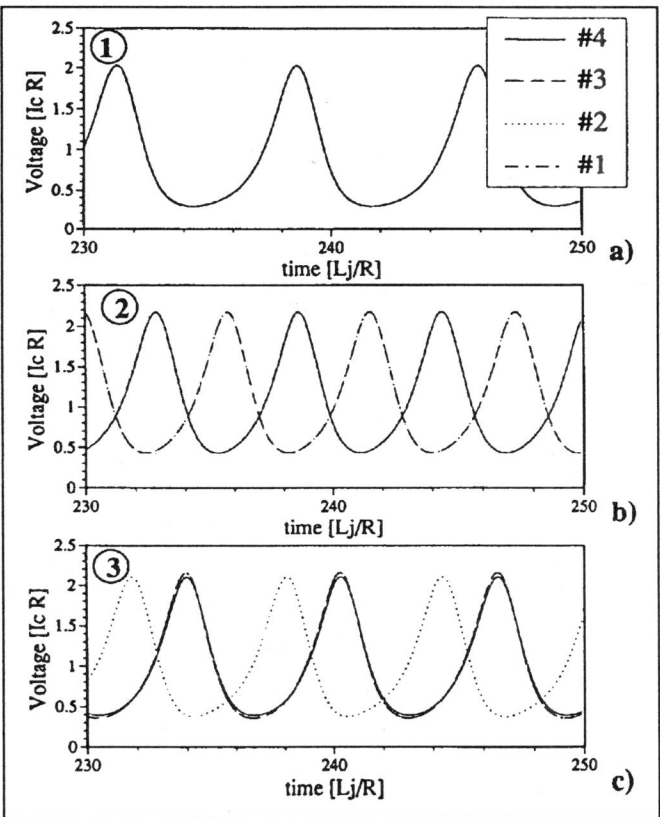

Figure 6. The waveforms of individual junction voltages for the states of Fig. 5. a) In-phase state. b) Junctions 1 & 4 in phase, junctions 2 & 3 in phase, but out of phase with (1 & 4). No power radiated. c) Junctions 1, 3 & 4 in phase, junction 2 "leads" a third of a period. Half the maximum power is radiated.

RADIATED POWER

As a measure of how good an array performs as an oscillator for a particular bias, we calculate the available radiation power. We are interested in power array would radiate broadside in the far-field. We define m^{th} harmonic power on unit $(1\ \Omega)$ resistance as

$$P(m) = (\Sigma v_j^{(m)})^2,$$

where $V_j^{(m)}$ is m^{th} harmonic voltage on j^{th} junction. This power is given in units of $(Ic\ R)^2$. We assume that the resistance R of the RSJC model (Fig. 2) includes both the radiation resistance and losses. The actual radiated power will at best be the power $P^{(m)}$ on resistance $\frac{R}{2}$.

Figure 5.b shows the normalized first harmonic power radiated in the broadside direction for different states of the array. The maximum power is obtained only in the in-phase states (points "1" and "4"). Significant amount of power is also obtained in states where most of the array works in phase, as is the case with state "2".

ARRAY PROPERTIES

Several important properties of arrays of Figure 3 can be derived from equations (5-10):

1.   In the absence of external magnetic field:

     - the UD array will be in phase for any DC bias; the in-phase state is a natural one for this array. Maximum power will be radiated at every operating frequency (Figure 5.b).

     - LR and CB arrays are symmetrical around the middle of the array; $m_{N-j} = -m_j$, which means that the junctions j and N-j are always in phase.

2.   The LR array is equivalent to the LL array with equivalent external magnetic field $\varphi_{ex}' = \varphi_{ex} - \frac{N}{2}a$.

3.   The larger the inductance parameter λ, the more in-phase states will be found in the given current bias span (eq. 8), and corresponding DC voltage and operating frequencies span. Similarly, the larger the array (N), the more identifiable "other" states will be found in the dV/dI-I curve (Figure 5.a).

MAGNETICALLY STEERABLE ARRAY

When an array is biased in the in-phase state ($i_{DC} = i_k$) the normalized relative time shift between every two neighboring junctions is the same and proportional to the external DC magnetic field:

$$\theta_j = \theta = \varphi_{ex} \bmod 1, \qquad \forall j \qquad (11)$$

This situation is shown in Figure 7. In Figure 7.a the LL array is biased at the in-phase state (point "4", Fig. 5.a) with no external magnetic field. When an external magnetic field equal to a quarter of the flux quantum is applied, the time shift between the voltages of every two neighbors is equal to a quarter of period.

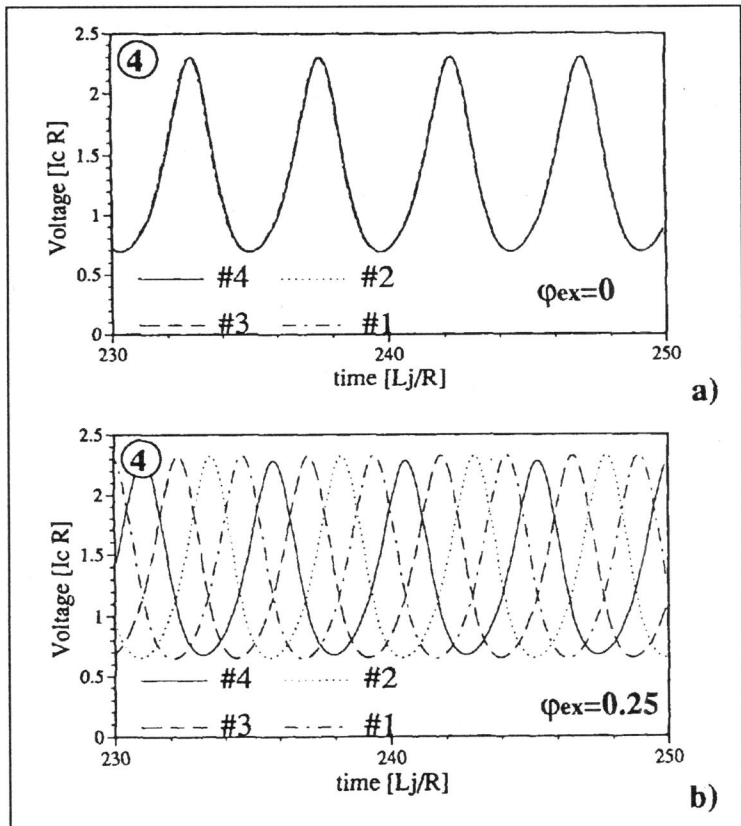

Figure 7. The waveforms of individual junction voltages for the in-phase state labeled "4" in Figure 5. a) No DC magnetic field supplied. b) Quarter of the flux quantum in every loop supplied by the external magnetic field. Voltage waveforms uniformly shifted by quarter of a period. The main beam of the radiation is steered from the broadside direction.

The time shift θT translates into the linear phase shift in the frequency domain $2\pi\theta$. Assuming that every junction drives one antenna, the quasioptical Josephson array becomes a phased array [9]. The angular position α_0 of the main beam in the H-plane far-field radiation pattern of the linear Josephson phased array becomes

$$\cos(\alpha_0) = -\frac{2\pi\theta}{\frac{2\pi}{\lambda_0}d_x} = -\frac{\theta}{d_x}\lambda_0, \tag{12}$$

where d_x is the spacing between the antennas and λ_0 is the free space wavelength. The broadside radiation corresponds to $\alpha_0 = 90°$ ($\theta = 0$). Equations (11,12) suggest that by changing the externally applied DC magnetic field φ_{ex} it is possible to steer the Josephson array radiation pattern in the H-plane. As stated earlier, for the LL, LR and CB biased array, this is only true if the array is biased in the in-phase state. Since the UD array is always in the in-phase state, it can be steered using DC magnetic field at any bias.

LIMITATIONS

The expressions (9,10) derived for the "other" states will hold only in certain range of array parameters and bias conditions. We have derived expressions (5) for the in-phase states starting from eq. (1) and assuming that all currents i_j are equal. These expressions always hold for the in-phase states. The same expressions (5) are found for "other" states if we solve eq. (1) with an assumption that DC currents $\langle i_j \rangle$ are approximately the same. The only part of the DC junction current that is different at every junction, due to DC self-field effects, is the supercurrent $\langle \sin(\phi_j) \rangle$. This part will be negligible if either the biasing current per junction is $\frac{i_{DC}}{N} \gg 1$ or if there is non-vanishing capacitance ($\beta > 1$).

In our account of DC self-field effects we assumed the noiseless environment and the identical junctions. Therefore, the stability of in-phase and "other" states to noise and variations in junction and array parameters remains to be further investigated.

STRONGLY COUPLED ARRAYS ($\lambda < 1$)

When the inductance parameter λ is small, it is evident from eq. (8) that the first in-phase state appears for very large DC biasing current, which translates into large DC voltage and operating frequency much above the critical frequency $\omega_c = (2e,)I_c R$. Depending on the capacitance parameter β and shunt resistance R, the operating frequency range is at best of the order of several ω_c. Therefore, the arrays with small inductances are operated in "other" states throughout the operating range. According to eq. (5), these states should be characterized with small time shifts between the junction phases/voltages. This is obvious, because in the limit of $\lambda \to 0$ the whole array operates as a single junction.

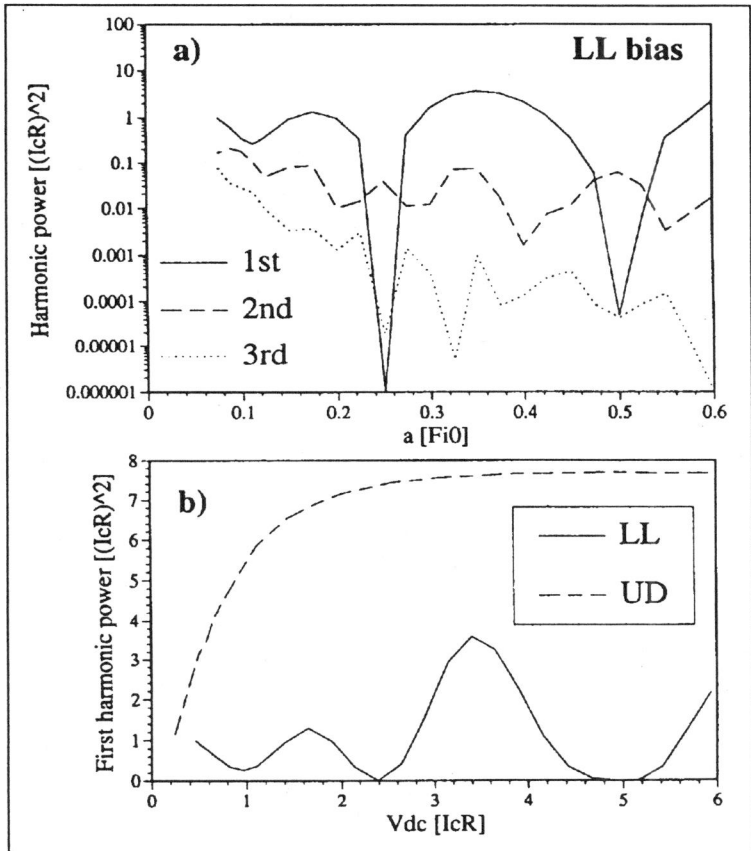

Figure 8. a) Harmonic power radiated by the 4-junction LL-biased array with $\lambda=\pi/5$ and $\beta=0.03$. Minimums correspond to odd number of half-flux-quantums in some of the loops. b) Comparison between the LL and UD array with some parameters. Varying amount of power is radiated in the very broad operating range, but never a maximum possible power, as in the case of UD array.

Figure 8.a presents simulations for 4 junction LL biased array with small inductance ($\lambda = 0.628$). The normalized harmonic radiated power is shown in the wide range of bias currents. The bias points a=0.25 and a=0.5 with no radiated first harmonic power correspond to the states where half of the junctions are in-phase and the other half out-of-phase, according to eq. (5.a). As seen from the Figure 8.b, the maximum first harmonic power is below that of the UD array, with all junctions in-phase. Figure 9 compares the first harmonic power of 4- and 5-junction array with same parameters. The 5-junction array shows additional minimums in radiated power corresponding to states a=0.125 k. These minimums occur whenever there is one or more loops of array occupied by odd number of half-flux-quantums.

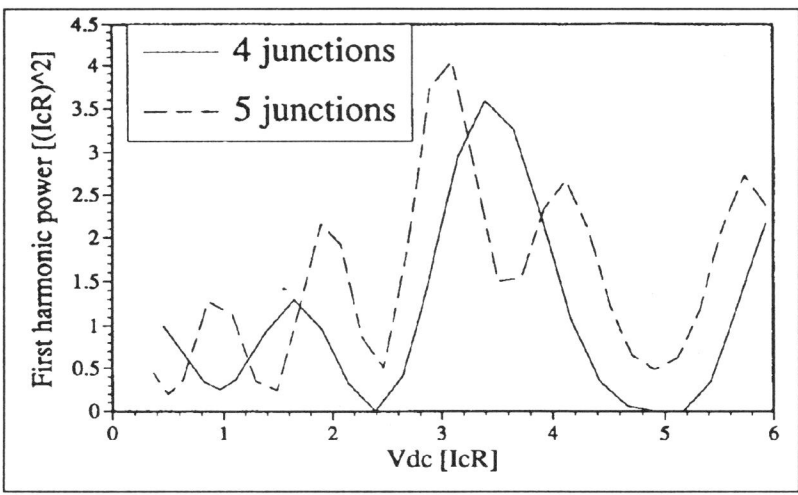

Figure 9. Comparison between the 4-junction and 5-junction small inductance arrays; $\lambda=\pi/5$ and $\beta=0.03$. As the number of junctions increase, more maximums and minimums appear throughout the operating range.

It is clear that in order for the array with small inductance to approach the performance of the UD array in the wide operating range, the condition for the inductance parameter must be $\lambda << \frac{1}{N}$. More precisely, if the total time shift across the array is required to be less than a quarter of a period, the condition is:

$$\lambda \; i_{DCmax} \leq \frac{\pi}{N+1},\qquad(13)$$

where i_{DCmax} is the DC bias at the end of the operating range. So, if we wanted a 4-junction LL biased array to approximately match the performance of UD array in Fig. 8.b, the inductance parameter should have been $\lambda = \frac{\pi}{120}$ instead of $\frac{\pi}{5}$. Such small inductances are rather unrealistic, specially because the λ parameter is proportional to the critical current I_C (eq. 1) which should be as large as possible for large output power.

As a final illustration, Figure 10 shows the influence of the Josephson junction capacitance. The capacitance does not influence the occurrence or existence of in-phase and "other" states. However, it has a severe impact on radiated power. Even at not very big capacitance ($\beta = 3$) the first harmonic power is decreased at least an order of magnitude and the operating range is reduced below $2\ \omega_C$ compared to the case of very small or no capacitance.

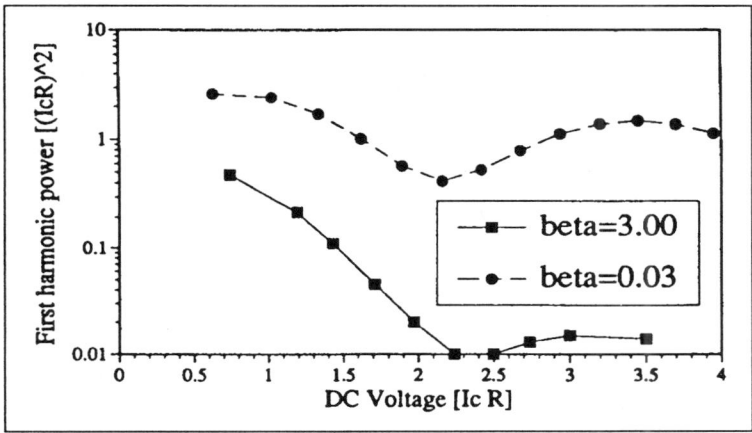

Figure 10.   Four junction LL biased array, $\lambda = \pi/10$: Influence of the Josephson junction capacitance on power output. As the capacitance is increased, the maximum power and the operating range rapidly decreases.

CONCLUSION

We have discussed how the DC biasing circuit determines the operation of linear parallel quasioptical Josephson junction arrays. We have shown that the maximum radiated power from the array can be achieved only at certain operating points, corresponding to the in-phase states. We have found that other states can be described by time-shifted phases and voltages of individual junctions, where the time-shift is determined from the DC biasing conditions. We have shown how the array can be steered from when in the in-phase state by application of DC magnetic field perpendicular to the array.

When the inductance parameter λ is large, there will be numerous in-phase bias points in the desired operating range. However, the stability of these states to noise and variations in junction parameters needs to be further investigated. When the inductance is relatively small, the radiated power will continuously change across the wide operating range, with several points where almost no power is radiated.

If one dimensional quasioptical arrays are designed, the UD biased array is a definite choice, because it is in the in-phase state at every bias point. The operation of this array need to be further analyzed when junction parameters are not identical. The extension of our considerations to 2-D arrays is straightforward, as long as rows of junctions are separately biased.

ACKNOWLEDGMENT

This work is supported in part by the Air Force Office of Scientific Research grant AFOSR-90-0233.

REFERENCES

[1] R. P. Robertazzi and R.A. Buhrman, "Josephson terahertz local oscillator," *IEEE Trans. Magn.* **25**, 1384-1387 (1989).

[2] K. Wan, A.K. Jain and J.E. Lukens, "Submillimeter wave generation using Josephson junction arrays," *Appl. Phys. Lett.* **54,** 1805-1807 (1989).

[3] S. P. Benz and C.J. Burroughs, "Coherent emission from two-dimensional Josephson junction arrays," *Appl. Phys. Lett.* **58**(19), 2162-2164 (1991).

[4] M. J. Wengler, A. Pance, B. Liu and R.E. Miller, "Quasioptical Josephson Oscillator," *IEEE Trans. Magn.* **27**(2), 2708-2711 (1991).

[5] T. Van Duzer and C.W. Turner, *Principles of Superconductive Devices and Circuits*, Elsevier, New York (1981).

[6] E. Ben-Jacob and Y. Imry, "Dynamics of the DC-SQUID," *J. Appl. Phys.* **52**(11), 6806-6815 (1981).

[7] W. H. Press, B.P. Flannery, S.A. Teukolsky and W.T. Vetterling, *Numerical Recipes in C: The Art of Scientific Computing*, Cambridge University Press, Cambridge (1988).

[8] J. Clarke and T.A. Fulton, "Origin of Low-Voltage Structure and Asymmetry in the I-V Characteristics of Multiply-Connected Superconducting Junctions," *J. Appl. Phys.* **40**(11), 4470-4476 (1969).

[9] B. D. Steinberg, *Principles of Aperture and Array System Design*, John Wiley & Sons, Inc., New York (1976).